T0178629

A Guide to Fire and Gas Detection Design in Hazardous Industries

In the last 15 years, the field of fire and gas mapping has grown extensively, yet very little is published on the subject. The text includes deeper discussions on important engineering factors associated with fire and gas detection, along with anecdotes and examples.

It will guide the readers on what to consider when you do not have access to proprietary standards, and how to interpret the design process even when one does not have access to a guidance document. The text covers important topics including visual flame detection, flame detection mapping, infrared point gas detector (IRPGD) operation, infrared open path gas detector (OPGD) operation, ultrasonic/acoustic design, and gas detection mapping.

The book plays the following roles:

- Explores practical aspects of designing a detection layout.
- Enables users in interpreting a detector data sheet and coverage analysis.
- Teaches readers working on a project to cut through the marketing of detection and to design an effective system.
- Inclusion of real-life experiences on projects will provide engineers with clear examples of where things can, and often do, go wrong.

It is an ideal text for professionals and graduate students working in the fields of occupational health and safety, fire protection engineering, and environmental safety.

The text discusses fundamental aspects of fire and gas mapping, which have been applied with great success in many parts of the world and are commonly adopted by the major operators in the process industries.

A Guide to Fire and Gas Detection Design in Hazardous Industries

James McNay

CRC Press
Taylor & Francis Group
Boca Raton London New York

CRC Press is an imprint of the
Taylor & Francis Group, an **informa** business

First edition published 2023
by CRC Press
6000 Broken Sound Parkway NW, Suite 300, Boca Raton, FL 33487–2742

and by CRC Press
4 Park Square, Milton Park, Abingdon, Oxon, OX14 4RN

CRC Press is an imprint of Taylor & Francis Group, LLC

© 2023 James McNay

ISBN: 978-1-032-16012-2 (hbk)
ISBN: 978-1-032-16014-6 (pbk)
ISBN: 978-1-003-24672-5 (ebk)

DOI: 10.1201/9781003246725

Typeset in Sabon
by Apex CoVantage, LLC

Not everything that is measurable is important,
and not everything that is important is measurable.
Daniel J. Siegel

Contents

Foreword

It was 3 pm, and I was working offshore in the warm waters of the South Atlantic Ocean. An immense bang occurred, and at that moment, the crew and I realised that something had gone seriously wrong. The general alarm wailed through the platform, and over 300 men and women made their way to muster. After sitting for eight hours in a humid mess hall, we found out that the operators had experienced the worst gas release in their history. Over 20 tonnes of natural gas had escaped via the overboard caissons and enveloped the platform. As the incumbent Fire and Gas Engineer, I watched the alarm matrix panel light up like Christmas day. In the days leading up to the event, I had noticed ice forming on the flare header, which was becoming so thick that it was blocking the path of one of the lines of sight of gas detectors. This ice should have been a warning to the process engineers aboard the platform, as it pointed to a blocked flare line caused by hydrate build-up. A process upset caused a small gas release, which, combined with the blocked flare header, resulted in the domino effect of a much larger incident—the classic breakdown of barriers in a Swiss cheese model, or as James would argue a breakdown of the system as a whole in catching this incident before the dominos started to fall. Thankfully, the gas did not find an ignition source that day, it dispersed naturally, and all the crew on-board went home safely to their families. In the weeks and months after this event, I realised that this was my career's 'why' moment. Why I do what I do, and why the design and engineering of fire and gas detection systems are crucial in moments like these. Fire and gas detection systems save lives and protect property, the environment, and reputation. It is for this reason that James' book is so essential for our industry. The need to design effective, safe, and reliable safety systems is as critical today as 50 years ago.

I first met James in 2010. I had a feeling, at that first moment, that he would do great things in our industry. In the 11 years that have passed, we have become close friends and have worked on many exciting fire and gas detection projects together. James recently completed his Ph.D. in Systemic Fire Safety Engineering while also leading the BSI Hazard Detection Mapping Committee as a chairman. James and I also developed the world-leading Fire and Gas Practitioner Programme, which is taught to safety professionals around the globe.

James' book, *A Guide to Fire and Gas Detection Design in Hazardous Industries*, is a fantastic tool to be utilised by safety and instrumentation professionals, enabling them to carry out robust fire and gas mapping assessments. For fire engineers, I recommend Chapters 4 and 5 comparing the nuances of differing flame detection technologies. In addition to this invaluable information, James has included the calculations required to equate the actual radiant heat output (RHO) of fires. This information, combined with example fire sizes for alarm and control actions and application-specific risk grades, enables designers to efficiently conduct flame detection mapping assessments. The combination of real-world knowledge, a performance-based design methodology, and the proper understanding of how a flame detector operates will only result in safe, repeatable, and cost-effective designs.

James' experience of leading Micropack and being involved in a diverse number of projects shines through in this book to provide a well-balanced, practical view on Fire and Gas Mapping. Its content includes, but is not limited to, design strategy selection, performance-based fire and gas mapping, selection of flame and gas detection technologies, and methodology and performance target examples. This book will prove an invaluable source of information to students, academics, engineers, and practitioners in the field of fire and gas safety.

Graham Duncan
Micropack (Engineering) Ltd.

Preface

In the 2010s, F&G mapping became a buzz term in the process industry. Many were under the impression that the field was a recent or emerging field, despite application of F&G mapping principles and tools dating back to the late 1980s. F&G mapping is not new, and robust methodologies were established in the 1980s, 1990s, and 2000s with the objective of applying repeatable, auditable, and uniform design through performance-based means. Contrary to popular belief, these approaches did not always result in excessive detection numbers through 'traditional prescriptive design' as has become the view of F&G design in the late 20th century.

The crucial point in the review of F&G systems is to ensure that the implementation of appropriate principles of design is not disrupted by the 'smoke and mirrors' which can be introduced to distract from gaps in knowledge or the application of an inappropriate design method. In the absence of any real volume of published, peer-reviewed journal papers on the subject, it is easy to simply accept this uncertainty with respect to F&G mapping and technology application. There are real dangers in applying F&G detection inadequately, and this must be addressed when the methodology is initially determined.

There are gaps in knowledge and understanding in F&G detection which this book will discuss. Some of these gaps in knowledge would require significant money and effort to address, neither of which appears to be forthcoming. While there will always be areas less understood in all fields of engineering, the approach to F&G design and technology has often appeared to demonstrate a desire to push away from discussing and acknowledging those murky areas in the attempt to promote a certain way of designing, or to promote the application of a 'tried and trusted' technology where it may not be appropriate. If there is a single objective of this book, it is to provide the reader with the tools to make an informed decision on what techniques and technologies are best suited to a specific application. It is unlikely that there will ever be a perfect solution in the grey area of F&G, but selecting the most appropriate approach and technology is critical in maximising the potential of achieving safety.

Along the spectrum of F&G designers, there are two extremes, of which most engineers sit somewhere in the middle. One extreme would advise that the sole requirement to design an adequate detection layout is a piece of software one can use with no prior knowledge. The second extreme is the practitioners who may 'overcomplicate' the field (in the eyes of the first extreme) in order to maintain their position as an industry expert rather than sacrificing their expertise to a piece of software. There has to be, in truth, a happy medium. In getting there, however, there has to be acknowledgement of the following points:

- The mapping software itself constitutes less than one-third of the overall process when one considers the role of risk-based targets, technology selection, mapping, and practical considerations of installation and commissioning. It therefore plays an important role, but it is not the sole driver of a sufficient solution.
- The role of a competent engineer, specifically in F&G, is important. This is not because the design of F&G is overly complex and onerous, but that it deserves the weight of accountability and subsequent respect which other engineering disciplines receive. It is important to remember that there are a number of key principles in application of these devices which, if missed, can have disastrous/expensive consequences.

Therefore, it would appear that the natural and reasonable assumption would be to ensure that an appropriate mapping tool is applied by competent F&G specialists. Even this, however, is not a fool-proof solution. All subsequently applied methods must be challenged to ensure that an appropriate design is achieved, and suitable referencing of all influencing factors must be provided. Importantly, these factors must also be correctly incorporated into the mapping tool.

There are common pitfalls which could result in a dangerous/inefficient detection design, and with the very nature of F&G design not being an exact science, those who fall into some of these pitfalls (selection of inappropriate detection technology, inappropriate detection location, inappropriate design methodology, etc.) may hide behind the aforementioned 'smoke and mirrors' of detection mapping. There may also be shortcomings from manufacturers/installers/integrators/maintenance teams, etc., in application and operation of the system, and as such, teething problems with an F&G system are extremely common. This is despite the fact that many of these issues could be picked up on paper at the beginning of the design process. This can also provide a direct financial saving to operators, as it is always cheaper to change something on paper than in the field.

With reference to the attempts at providing independent guidance on the matter, ISA TR84.00.07 Guidance on the Evaluation of Fire and Gas System Effectiveness (1) provides the appropriate starting point of a design basis and intentionally allows the user to apply differing methodologies. This may, however, give rise to those not familiar with F&G design applying an inappropriate methodology based on, for example, a simplified version of mapping which is more easily comprehended but may not be appropriate in the given circumstance. Conversely, it can be applied by designers trying to force a methodology which works for other safety systems with which they are more familiar, and then justifying this as compliance with an international guidance document on what is a very unique safety system.

With the increasing commercialisation of F&G mapping packages, the ability to carry out the mapping process is being spread across multiple disciplines. It is therefore of paramount importance that if the selected method of design requires application of such tools, the mapping tool being applied is appropriate to the application in hand, and has been designed/incorporated by personnel with experience in F&G design. Adequate competence must also be demonstrated by the user, through qualification, experience, or appropriate training (on methodologies, software operation, and the practicalities associated with F&G hardware).

Whichever methodology is applied, it is crucial that parties involved on both sides of the project (designers and implementers) are happy with the methodology at kick-off, are fully aware of the strengths and limitations of the selected methodology, and work together to ensure that this is appropriate for the specific application.

This book aims to discuss the fundamental aspects of F&G mapping, which have been applied with great success in many parts of the world and are commonly adopted by the major operators in the process industries.

This book shall also cover the lessons learned from project experience in designing F&G layouts. The philosopher George Santayana put it, *'those who cannot remember the past are doomed to repeat it'*. Perhaps more appropriate to F&G design was his quote *'Theory helps us to bear our ignorance of fact'*.

Some concepts, in this book, are embedded deep within the theory of detection operation and risk analysis which can be challenging to apply to a real-life situation on a plant, and are also, let's be honest, rather dry. I have therefore endeavoured to include some 'real-world' experiences into the book to shed light on how decisions made in an office block hundreds, if not thousands, of miles from the facility can have a significant impact. These hypotheticals will, I'm sure, be all too real to those in the world of safety engineering.

Each chapter will begin with snippets from hypothetical meetings/situations which, as previously mentioned, those in the safety industry will likely find familiar. It is important to note that, while these are based on my experience in the safety industry, these are not verbatim stories. These stories, rather, serve the purpose of putting the content of the book in context of reality, and my hope is that it makes the content more easily understandable and applicable to the real world. My ultimate goal is that this will help designers to notice such situations as they are occurring, and will also increase the likelihood that we can move away from such scenarios. I fear that the nature of the tales are a sign of a systemic failure to learn from history—a particularly problematic predicament in an industry susceptible to boom and bust . . . so to speak.

Acknowledgements

Special thanks go to Micropack (Engineering) Ltd., as an organisation which granted permission to utilise many of the F&G-related lessons learned over the course of the last three decades. Use of the Micropack Test Facility in Portlethen, Scotland, has also been of unquantifiable benefit, so special thanks go to every member of staff for their tireless assistance and expertise.

Thanks also go to FABIG, BSI, ISFEH, Draeger, and Honeywell, whose permissions to include some valuable content are greatly appreciated.

A special acknowledgement goes to my colleagues on the working group who developed BS60080:2020. While early in its publication, I believed that this document will have a special contribution to the improvement of F&G detection layouts across the UK, and beyond. Special mention to Samer Bachir, Bruce Hopwood, Tim Jones, Doug Longstaff, David Orr, and Randall Williams for tireless efforts in generating the content. Thanks also to Paul Cuddeford, our BSI Editor, who survived the many long days of editing in order to get it over the line. While there is always scope for improvement and expansion, the release of this guide felt like a watershed moment in the design of performance-based F&G mapping.

Acronyms and Abbreviations

2D—Two-Dimensional
3D—Three-Dimensional
AEGL—Acute Exposure Level Guidelines
ALARP—As Low as Reasonably Practicable
ASET—Available Safe Escape Time
BLEVE—Boiling Liquid Expanding Vapour Explosion
BSI—British Standards Institution
BTC—Baku Tbilisi Ceyhan
CAPEX—Capital Expenditure
CCTV—Closed Circuit Television
CFD—Computational Fluid Dynamics
CO_2—Carbon Dioxide
CPS—Counts per Second
DNV—Der Norske Veritas
EMSA—European Maritime Safety Agency
ERPG—Emergency Response Planning Guidelines
F&G—Fire and Gas
FABIG—Fire and Blast Information Group
FM—Factory Mutual
FOV—Field of View
FPSO—Floating Production Storage Offloading
FSE—Fire Safety Engineering
H_2S—Hydrogen Sulphide
HCR—Hydrocarbon Release
HRR—Heat Release Rate
HSE—Health and Safety Executive
HVAC—Heating Ventilation Air Conditioning
IChemE—Institute of Chemical Engineers
IEC—International Electrotechnical Commission
IMO—International Maritime Organisation
IR—Infrared
IRPGD—Infrared Point Gas Detector
ISA—International Society of Automation

ISFEH—International Seminar of Fire and Explosion Hazards
JIP—Joint Industry Project
K—Kelvin
KHz—Kilohertz
kW—Kilowatt
LEL—Lower Explosive Limit
LFL—Lower Flammable Limit
MAH—Major Accident Hazard
NAE—National Academy of Engineering
O&M—Operations and Maintenance
OFD—Optical Flame Detector
OPEX—Operating Expenditure
OPGD—Open Path Gas Detector
ppm—Parts per million
PT—Performance Target
QRA—Quantitative Risk Assessment
RHO—Radiant Heat Output
RRF—Risk Reduction Factor
RSET—Required Safe Escape Time
SCAT—Systemic Causal Analysis Technique
SFPE—Society of Fire Protection Engineering
SHORE—Systemic HAZID and Operational Risk Evaluation
SIL—Safety Integrity Level
SIS—Safety Instrumented System
SPL—Sound Pressure Level
STEL—Short-Term Exposure Limit
STPA—Systems Theoretic Process Analysis
TGC—Target Gas Cloud
TR—Technical Report
TWA—Time Weighted Average
UK—United Kingdom
UKCS—United Kingdom Continental Shelf
UKOOA—United Kingdom Offshore Operators Association
UV—Ultraviolet
VFD—Visual Flame Detection
μm—Micron

About the Author

James McNay, Ph.D., has committed his career to the improvement of fire safety in both the built environment and the high-hazard industries. James is currently Director of Fire Safety (Scotland) within the Fire Division of Hydrock working within the Built and Natural Environment. Previously, he served as Managing Director at Micropack (Engineering) Ltd. and specialised in the design of fire and gas detection systems in high-hazard industries. Both as a practicing specialist detection consultant, and while managing the engineering and consultancy teams at Micropack Headquarters in Aberdeen, Scotland, James had significant involvement in the design of F&G detection systems in the Gulf of Mexico, Alaska, North Sea, Caspian Sea, Australia, Qatar, Oman, Malaysia, Thailand, and Vietnam.

Through this experience, James has been involved in the generation and development of some of the most widely adopted F&G standards applied across the world by various hazardous facility operators. Most notably, James chaired the BSI Committee tasked with generating BS60080 Explosive and Toxic Atmospheres: Hazard Detection Mapping: Guidance on the placement of permanently installed detection devices using software tools and other techniques in 2020.

James has also served as Fire and Gas Chair for the Safety and Cybersecurity Division of the International Society for Automation (ISA) between 2015 and 2020.

James received his Ph.D. from the Department of Naval Architecture and Marine Engineering at the University of Strathclyde, specialising in systemic safety engineering. In the development of 'A structured, systemic methodology to improve maritime fire safety in machinery spaces', James aimed to understand and address why industry often witnesses the same accidents repeat themselves. He discovered that when addressing the fire safety problem from a systems' perspective in both design and operation, obvious opportunities for improvement, which are often and easily overlooked, become evident. Notably, this finding is not solely applicable to industrial fire safety and has provided James the desire to address challenges from a systemic perspective, from complex technical issues on a high-hazard facility or high-rise building, to everyday problems like how to prevent his much beloved dog pulling excessively on the lead on what should be a relaxing walk through the Scottish countryside.

1 Introduction to Fire and Gas Detection

Potential for disaster is inherent within the very nature of the high-hazard industries. Breaks in containment, with subsequent ignition, can result in significant fire and explosions. Historically, occurrences of this are well documented. Such events can result in vastly different outcomes, presenting various escalation effects and resulting damage.

In 1988, the Piper Alpha disaster in the UKCS gave the oil and gas industry a wakeup call to the potential for disaster offshore, although anecdotal stories would suggest that the dangers were being seen in the lead up to disaster. A breakthrough from this event was the increase in awareness of the importance of safety and safe working practices. Following the disaster, the oil and gas industry invested great time, money, and effort in the development of new technologies, or the improvement of existing technologies, and the improvement of overall system safety and working practices/procedures. This investment was in the effort to ensure that such a disaster of the scale of Piper Alpha would never happen again. One such safety system which was given special attention was that of F&G detection.

Following the developmental improvements in detection technology, the decision-making process on where to position these detectors offered room for improvement. The focus on placing detectors based on the specific area's hazards, intended to be mitigated, was the driver behind performance-based F&G mapping.

The result of this increased focus on detection mapping has given rise to various processes to 'map' detectors in the last few decades, with increased focus particularly in the last decade. This process of Fire and Gas Mapping is over 30 years old, and is not as recent a development as some would believe.

This application of F&G mapping, however, and the lessons learned from Piper Alpha have not resulted in the elimination of major accidents in the hazardous industries, noting the comparatively recent Deepwater Horizon explosion and subsequent environmental disaster. While a state-of-the-art F&G system may not have been able to prevent the event from being as significant as it was, it shows the importance of being able to take successful action before the situation gets out of control.

DOI: 10.1201/9781003246725-1

A break in containment can result in a variety of scenarios and can take a number of different forms, which can cause significant damage to a facility, and beyond. Take the example of a Boiling Liquid Expanding Vapour Explosion (BLEVE). Such an event requires the correct properties of a vessel containing liquid and gas, coupled with heating from an external fire source. Under the correct conditions, this sustained heating can result in the boiling off of the liquid in the vessel, resulting in the unexpected containment of a now high-pressure flammable gas. The vessel in question is not designed to withstand such a pressure and the resulting break in containment can be catastrophic.

This is an extreme example of escalation, but the pressure and composition of a process pipe, vessel, or tank will have a significant impact on the risk posed from a break in containment, or secondary fire impacting that equipment. Potential hazards can range from gas/liquid spray fires, flash fires, fire balls, pool fires, etc.

Considering the release of flammable gas, this can present an explosion hazard within the area of release if sufficient congestion exists, can present an explosion hazard to adjacent areas, and can also present a risk to non-hazardous areas where the conditions allow for the migration of the gas across, and potentially out of, the site. The same is also true, although the specific hazard is different, for toxic gas releases.

This demonstrates the importance of applying the most appropriate methodology with an adequate understanding of the specific hazards at hand to ensure that a fire or gas release is detected as soon as required in mitigating the event.

It is not solely the design methodology which is critical, however. The understanding of the available hardware is critical in designing an adequate system. As each fire and gas detection technology is unique in a sometimes subtle, sometimes significant, manner, it is critical that technology is selected on the basis of suitability to both the hazard and the environment.

This book will demonstrate that there is no such thing as a perfect detector; therefore, any limitations of a technology which are understood and accepted must be documented. This allows for a suitable audit trail of decision making, and allows for a continual improvement process, as new technologies become available, or the hazards on site develop through the facility lifecycle.

While application of theoretical calculations of modelling accuracy, or calculations of probability of failure on demand based on leak frequency and likelihood of escalation are important in safety engineering, they may not be adequate. There exists a systemic potential for failure in selecting inappropriate technology, or failing to model the capability of technology during design which will undermine any quantitative analysis. This was the basis of the British Standard BS60080 (1) in encouraging that practicality in detection application is not sacrificed by focusing solely on mapping software and potentially misleading quantitative failure calculations.

The potential for disaster is all too apparent within any hazardous industry if a holistic and systemic approach to safety is not applied in design and operation. This is certainly the case with F&G detection systems.

The move in Oil and Gas towards elimination of the potential for 'fail to danger' of safety systems, with the development of IEC 61508 and IEC 61511, shows the importance of a functional and reliable F&G system.

Whether a fire and gas detection system falls under the remit of functional safety guidance such as IEC 61508/11 (2, 3), and subsequently be classed as a SIS, is out of the scope of this book. Such a field has a wealth of available books (4, 5); however, this book will focus solely on the critical elements of design of F&G layouts and technology application. Regardless of one's stance on the issue of the F&G relationship with SIS, the issue of reducing the potential for 'fail to danger' through practical design is one which must be addressed.

This alludes to a philosophical and practical question; does a completed F&G mapping model equate to an adequate demonstration of detection adequacy? This book will evaluate the current methods of dealing with both mapping and the application of detection technology to shed light on the wide range of applications and methods which can be applied. The author hopes that, through a holistic evaluation of F&G mapping techniques and technology, the reader will be best placed to understand what is needed from the F&G system in design and operation. It is also hoped that, when an approach is being presented, which is not in the best interests of achieving safety in the field, the reviewer will be equipped with the knowledge to challenge such designs.

If there are two overarching philosophies behind the authoring of this book, they are as follows:

1) Best practice/design standards should influence what a software calculates and presents, not the other way around. Software is a tool to assist a competent design engineer to verify a design. It pains me to consider design standards moving away from well-established and trusted practical design methods, solely on the basis of using a software feature presenting solely aesthetic gains.

2) Connected to the first philosophy mentioned earlier, to quote Daniel J. Siegel: 'not everything that is measurable is important, and not everything that is important is measurable' (6).

The book will begin with an overview and introduction to performance-based F&G design and is followed by a discussion on systemic safety engineering. While systemic safety engineering is not well established in the F&G detection world, it is certainly the trajectory being taken in the safety industry. The sooner engineers can begin to implement systemic safety engineering in their processes, the more effective our safety systems and processes will become.

The book will then discuss flame detection technologies, discussing the technologies available at the time of writing along with the strengths and limitations of each.

This is then followed by a detailed review of flame detection mapping. This shows how the knowledge of the technologies, the hazards in question, the environment, the software tools applied (if necessary), and more, all contribute to designing a detection layout which can be deemed adequate.

Moving on from flame detection, the book will then discuss flammable gas detection technology, discussing the prevalent technologies available at the time of writing, with their strengths and limitations. The technology discussion focuses heavily on flammable gas detection technologies, and only briefly discusses technologies suited for toxic gas detection, and specialist technologies such as ultrasonic/acoustic gas detection.

As with flame detection, a detailed discussion on flammable gas mapping is then presented.

While the aforementioned topics constitute the primary content of the book, additional notes and discussion are presented on specialised hazards such as HVAC detection, ultrasonic detection design, and toxic gas detection design, closing with a brief discussion on competence. Being a specialised field, competence in designing those F&G systems is critical but unfortunately not well defined.

It is the author's hope that this book can be of assistance to those first embarking on increasing their awareness of F&G detection systems, but also those who have spent a career in the industry. It is also hoped that the book will provide assistance to those designing, auditing, approving, or maintaining systems. If the book can spark a query here, or a clarification there which leads to an opportunity for improvement, it will have succeeded in its objective.

In a field where the adequate level of safety is not binary, put yourself in the shoes of those at risk before making a decision.

References

1. BSI. BS60080 Explosive and Toxic Atmospheres: Hazard Detection Mapping— Guidance on the Placement of Permanently Installed Flame and Gas Detection Devices Using Software Tools and Other Techniques. BSI Standards Limited; 2020.
2. IEC. IEC 61508 Functional Safety of Electrical/Electronic/Programmable Electronic Safety-Related Systems. Geneva, Switzerland: IEC; 2010.
3. IEC. IEC 61511 Functional Safety—Safety Instrumented Systems for the Process Industry Sector. Geneva, Switzerland: IEC; 2017.
4. Beurden I, Goble W. Safety Instrumented System Design: Techniques and Design Verification. ISA; 2017.
5. Scharpf E, Thomas H, Stauffer T. Practical SIL Target Selection—Risk Analysis Per the IEC 61511 Safety Lifecycle. 2nd ed. Exida; 2016.
6. Siegel D. Mind: A Journey to the Heart of Being Human. Illustrated ed. W. W. Norton & Company; 2016.

2 Performance-Based F&G Mapping

There are a multitude of factors to incorporate in an adequate F&G system. In order to effectively account for these in an unpredictable environment, the designer must apply a method of design which can account for the hazards, the environment, and the technology available. The designer also must be able to demonstrate that such factors have been accounted for. One such method is through the application of performance-based F&G mapping.

Application of performance-based methods in engineered design can, however, lead to some interesting behaviours. Application of the method alone is not a suitable solution.

The leading anecdote in this book demonstrates the potential for dubious application of performance-based design, and shows that intent during design is critical. It presents a scenario where the quantitative analysis of event probability, common in safety engineering, is applied to dilute the requirement for mitigation measures.

In this hypothetical case, a detection design is created which meets predetermined and previously agreed detection targets. The resulting number of detectors, however, is deemed by the facility owner as 'too high' (despite the analysis showing that the proposed solution is the optimal number to meet the targets previously set by the operator). Said operator hatches a plan and sets the safety engineers to work—implement a risk-reduction criterion for each detector to justify its addition based on probability of loss. What results is the addition of events leading up to the initial break in containment, including analysis of leak likelihood, reduced maintenance, hardware availability, and reliability factors. This then presents an overall risk value where likelihood and consequence are calculated and verified against an acceptable ALARP value. This type of analysis is standard practice in safety engineering (whether adequate or not), and, in light of more recent understanding of accident causation, it is in fact somewhat outdated (1–4). Regardless of adequacy, in this example, the approach should have been agreed at the start of the process, not introduced at the end of the process after the design is complete. This 'moving of the goal posts' shows that the intent of the process may not be to achieve an adequately safe system.

DOI: 10.1201/9781003246725-2

The process discussed applies techniques such as event tree analysis to determine the risk of certain outcomes, and subsequently determine if the risk is tolerable. In theory, this then shows where additional detectors are required to reduce the overall risk to ALARP. In this instance, as additional 'events' are added to the event tree, with associated 'likelihood' being <1.0, the final risk value of release (the initiating event for the F&G system) is lowered, therefore (arguably artificially) reducing the risk reduction achieved by the F&G detection. Hey presto, the resulting risk reduction achieved by each individual F&G detector in isolation becomes negligible, as the likelihood of event has been reduced across the board. No additional detection is required. A happy facility owner, but has the system really been demonstrated to achieve an adequate level of safety?

The approach discussed here would apply established safety engineering principles, and the purpose of this book is not to condemn the established field of safety engineering. Quite the opposite, this anecdote shows that *intent* behind application is critical to eliminate systemic causal factors creating the creep towards disaster. If a process is followed and the answer is not desirable, one should not continue to move the goal posts until the desired answer is achieved. If this is the case, what was the point in the process? This is clearly a systemic route to failure through application of quantitative risk analysis with a predetermined agenda.

This critique, which is attributed to quantitative risk assessment (QRA) in general, is the slippery assumption of event independence of basic events in fault or event trees, as demonstrated by Wikman (5). Should this assumption turn out to be false? It would pave the way for common-cause failures (6). A review of 609 aviation accidents found that at least 11% of them were caused by common-cause failures (7). There is no basis to assume a lower rate in the hazardous industries, given lower levels of standardisation and widespread application of F&G mapping tools which propose to design the system automatically with minimal engineer intervention. The assumption of independence can also lead to so-called Titanic coincidence (8), when many low probability events are multiplied to estimate the top event probability. As with the anecdote, this results in a negligible probability of the top event, which is then discarded as impossible or cost-ineffective to attend to.

As generally applied in the established field of Fire Safety Engineering, it would be wise when designing *mitigation* systems to start from the probability of fire as 1.0. As prevention and mitigation are separate sides of the same coin, they are clearly connected, but a truly safe system is achieved and maintained through resilience of both. Reliability of one should not be discarded due to perceived infallibility of the other. This will be further elaborated in Chapter 3 when systemic safety engineering is discussed.

Methodology

For the sake of clarity, this book's focus will be on performance-based methods of F&G design. Prescriptive-based methods require little attention, as these, by their very nature, are not challenging in their implementation. Where an application is simple, or the hazards presented are relatively of low risk [and where a suitable design criterion exists which can be easily followed, such as BS5839 (9)], prescriptive designs often take precedence. Where this book will focus on is the forum where the prescriptive approach is not applicable and a deeper consideration and understanding of the hazards, and how they are to be detected, are required.

While an overview of the specific performance-based techniques is not provided in this book (with this being covered in official standards, such as BS60080), the fundamental philosophy behind F&G mapping will be examined to assist in decision making at all stages of design and technology selection. This will also assist in diverting from the tired old 'this method is better than that method' type arguments often presented in the literature.

Performance-based design requires a target or a goal to be set, against which the design can be compared. The process sets a target which provides the equivalent level of safety against an established or 'known to be safe' level, which the operator and designer can demonstrate that the design will achieve. This 'comparatively safe' criterion becomes a challenge with F&G detection design in a hazardous industry, as the comparable standards are typically related to the internal application of smoke or heat detection in the built environment.

The designer therefore has to review the hazards and overall risk in the area and determine what the transition point from an acceptable, to an unacceptable fire or accumulation of gas would be, and then has to set this as a 'Performance Target' (PT) of the detection system. These targets would always require to be accepted by the owner of the facility as ultimately, and they will have responsibility for the risk and should sign off on the acceptable loss criteria.

There are various objectives in setting performance targets for the F&G system. The performance targets should record the hazards present on the facility and the associated damage they can cause. Based on this escalation potential and tolerable level of damage, the operator's expectation of performance of the detection system is specified.

This system performance (including the layout, technology, and mitigation actions) must be clearly communicated to the system designers and those implementing it.

The performance targets, crucially, must also provide an acceptability metric by which they can be audited and verified through the lifecycle of the facility.

In essence, the process of performance-based F&G mapping must question what the escalation potential in the area is based on the specific hazards, what consequence level is to be prevented, and what we expect the F&G detection system to contribute in mitigating the escalation potential.

The result of the approach will be the definition of the minimum extent of a hazard, which can occur where a detector will be expected to respond (for example, the target fire size at which the flame detectors would be expected to respond, or the minimum target gas cloud in which gas detectors would be expected to detect). This approach should apply the knowledge and experience of the operator along with the expertise of the designer.

It is noteworthy that the specified target would often not be the *expected* event which can occur. This is a common misunderstanding in the application of fire hazard analysis or gas dispersion analysis in detection design. It is common for people to believe that these studies present the events which can occur, and then to dangerously jump to the conclusion that these should therefore become the target for our F&G detection system. This is often (but not always) untrue.

Consider a facility where the worst-case scenario fire is determined to be a 9-m-diameter pool fire, which is subsequently set as the target for flame detection. It is reasonable to assume that no fire smaller than this 9 m pool fire could be reliably detected. It may be detected, but as the design is to the 9 m diameter, anything smaller is not expected to be detected.

What level of mitigation is therefore achieved by the detection system which is designed to respond only when the worst-case scenario fire occurs? It may be the case that the detection is there only to make the operator aware that the worst-case scenario fire has occurred, but this would need to be documented in the performance targets. It is important to remember that the intention of detection is normally to detect smaller fires (either from the equipment of concern or from adjacent fires impinging on the equipment in question) in order to prevent them escalating to the worst-case scenario fire.

Additionally, the designer must consider and acknowledge that flame detectors are not designed or certified to detect such large fires. This is a critical, and often overlooked, issue in design which will be further investigated in Chapters 4 and 5.

Consider again the facility with the large target fire size (the aforementioned 9-m-diameter pool fire). Designing to such a large fire will ultimately reduce the required number of flame detectors. This is a result of the visibility of such a fire being clear from any location on the facility.

In reality, however, there may be a smaller outbreak of fire in the vicinity of the equipment of concern, which results in a controllable fire similar to the 1 ft^2 pan fire. This fire could be the event which results in the failure of the equipment of concern and the resultant 9 m pool fire, but it will likely remain undetected.

Consider also a low-risk piece of equipment where the worst-case scenario fire is the 1 ft^2 pan fire. Where designers design to this fire, the engineer would be required to add a potentially significant number of detectors to see this fire due to its small size. Why would a designer add extra flame detectors when the risk in the area is so low as to only result in a small fire? The short answer is they generally should not. This demonstrates the criticality of adopting an approach which philosophically, fundamentally, and practically complies with the nature of risk-based F&G detection. Always consider the fundamentals of what the detection system is designed for, and what is intended to be achieved.

Prescriptive Versus Performance-Based Design

A train of thought exists within F&G mapping that a design will fall into either prescriptive or performance-based design. While this is naturally the case, the specifics are a little more complex. As previously mentioned, one such engineering discipline which F&G mapping can take its lead from is Fire Safety Engineering (FSE) in the built environment.

Within FSE, there are prescriptive designs and performance-based designs. Prescriptive designs will apply code-based rules within design guides or standards such as Approved Document B in England (10), and the Building Standards Technical Handbook in Scotland (11), for example. Although these documents are prescriptive in their nature, they are only practical guidance on how engineers can comply with the Building Regulations. Design which applies the guidance of Approved Document B, for example, is not guaranteed to comply with the performance-based building regulations. The Approved Documents cannot account for all building types and situations. Where this book references the guides as 'prescriptive', it is in the sense of the prescriptive statements provided, i.e. the provision of a maximum travel distance. This is not to suggest that the Approved Document B, for example, is a prescriptive design guide which guarantees compliance with the Building Regulations. It is not.

Where designs do not, or cannot, achieve the specifications in these documents, a performance-based approach is applied. The level of detail, however, varies considerably based on numerous factors.

If the design contains a minor deviation from the codes, a simple engineering evaluation on the deviation, and review of the corresponding safety factors designed into the building, will suffice. One such example would be where the maximum travel distance for escape is exceeded in a building by 5 m. When looking at the occupancy of the building and assigning a walking speed of 1.2 m/s, this would equate to approximately an extra 4 seconds added to escape time. Perhaps the building also applies a smoke detection and alarm system which is not specified in the codes for such a facility and is therefore a fire safety enhancement which will reduce the pre-alarm time of evacuation. A fire engineer can look at these factors and justify that an equivalent level of safety is achieved in the building and sign off the design. Such a simple deviation would not require a complex numerical analysis, but is still technically performance based.

Compare this, however, to the design of a state-of-the-art high-rise building with multiple occupancies and inherent unique architectural features where the prescriptive codes are simply not applicable. A building like this would likely require a complex fire-engineered design and likely a quantitative analysis of various factors including evacuation time, smoke development and spread through the building, and enclosure fire dynamic calculations to verify structural integrity requirements.

All of these would work towards gaining a complete picture of the building's fire safety and determining if that level of safety is adequate.

The situation is more complex still, as with the performance-based approach, we have the facets of both deterministic analysis and probabilistic analysis, with absolute and comparative acceptance criteria.

Deterministic analysis evaluates various parameters in absolute terms. An example from a fire safety engineering approach would be 'is the smoke layer in a compartment fire capable of descending below 2.5 m based on the room geometry, fire load, and ventilation conditions?'

Probabilistic analysis calculates the probability of an event or consequence. A similar example to that given for deterministic analysis would be 'what is the likelihood of the smoke layer descending below 2.5 m?'

How this relates to performance-based F&G mapping should hopefully become apparent by the end of this book, but I will begin to paint the picture here. While the comparison between deterministic analyses in an FSE perspective is not identical to that in F&G mapping, the concepts are similar in the sense that it is a numerical approach to calculating if a condition of concern will occur, i.e. will a 5 m spherical cloud of gas remain undetected? This is not to suggest that a spherical cloud is anticipated or even realistic. Established enclosure fire dynamics recognises that a perfect plume in the centre of a room is not how a fire behaves in reality, but engineering calculations are based on such assumptions. These are then interpreted by fire engineers. The same is required in performance-based F&G mapping.

The concept of proportionality is critical in hazardous area's F&G mapping and correlates with the FSE approach. When looking at a hazardous facility, we can begin to consider facilities which would fall under the same categories as those discussed previously.

A basic internal low-risk fuel storage facility, for example, may fall under a prescriptive routine. An offshore Oil and Gas platform, however, is likely going to require a performance-based approach. This is not simply black and white, however. Which performance-based approach should one take?

This will depend on the area being reviewed. External storage of diesel tote tanks, for example, may only require an analysis similar to the travel distance issue presented earlier. 2D plot plan detector placement based on good engineering judgement can be justified without further numerical analysis. When looking at primary hydrocarbon processing areas, however, some numerical analysis is almost certainly required.

For an area like this, a performance-based, deterministic approach could be justified. Does a damaging volume of gas occur here, and if it does, will it be detected? Such an approach would assume that gas release is credible and is therefore assumed to have occurred. This becomes the starting point for the analysis. The designer will set the criteria for detection and ensure that this is met. This reflects closely to the volumetric-based design which will be discussed later in this book (sometimes also referred to as the geographic method). This method, however, is often referred to as prescriptive. Indeed, if the method is blindly applied as '5 m spacing regardless of the application', this would be prescriptive, but this would be a misapplication of the methodology. The volumetric approach is inherently performance-based, but more closely resembles deterministic performance-based design in the build environment.

As we move towards facilities which are ultra-high hazard and complex in both layout and processing, with more complex safety measures, probabilistic performance-based design may be applicable. Such facilities processing hydrogen at high pressures, for example, may benefit from application of numerical dispersion and explosion analysis tools. This may be because there is no enough evidence from previously developed designs which demonstrate safety from which to base a deterministic design. The volumetric approach can, however, still be relevant in these applications, just as the scenario-based approach can be applicable in some circumstances which do not present as high a hazard. Ultimately, however, the decision should be based on using the most appropriate method for the specific application, as agreed with the facility owner based on their approach to risk.

Figure 2.1 represents a simplified version of the selection of a design strategy.

Table 2.1 shows examples of gas detection acceptance criteria, or performance targets, for a gas detection design applying deterministic or probabilistic strategies, with the corresponding absolute and comparative acceptable targets.

Figure 2.1 and Table 2.1 are adapted from BS 7974 (12) examples of methodology selection and acceptability criteria.

Mitigation Action

Regardless of the method applied, the means of mitigation is critical. Success of the detection of the hazard of concern is only useful if an adequate action is implemented upon successful detection.

The mitigation action is the implementation of an action which will inhibit the continuation along the path towards an escalated and uncontrollable state. Put in a simple real-world context, you may have a fire alarm which detects a fire in a kitchen and produces an audible alarm. From the noise of the alarm, an occupant proceeds to the kitchen and places a fire blanket over the pot which has caught fire. The mitigation here is the occupant

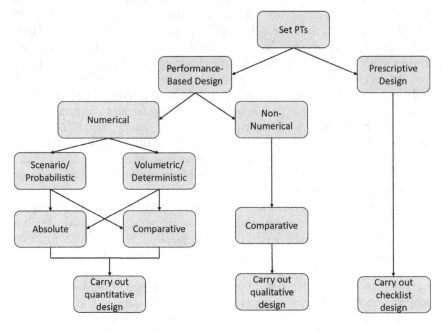

Figure 2.1 Design strategy selection.

Table 2.1 Methodology and Performance Target Examples

Performance-Based Method	Performance Target
Numerical Volumetric/Deterministic	
Calculate the volume of gas which would present an explosion overpressure of >150 mBar	**Absolute:** Gas cloud with diameter greater than that cloud size shall be detected by at least one gas detector
Review the area to calculate the volume where gas clouds would remain undetected	**Comparative:** Target gas clouds of a directly comparable acceptable design to be detected by at least one gas detector
Numerical Scenario/Probabilistic	
Assess the probability of a dangerous gas cloud remaining undetected	**Absolute:** Probability of gas cloud remaining undetected <10%
Assess the probability of a dangerous gas cloud remaining undetected in a directly comparable application which has been classed as acceptable	**Comparative:** Probability of gas cloud remaining undetected to be the same or less than the comparative acceptable design
Non-Numerical Qualitative	
Minor deviations exist from a previously acceptable design solution (e.g. 5 m detector spacing).	**Comparative:** Engineering judgement based on the facility risk and additional safety features to be considered to determine adequacy against a typically 'safe' design.
Prescriptive	
Assess the detection against the prescriptive requirement	Check compliance/non-compliance

intervention. This mitigation has been allowed to be successful from the audible alarm, the presence of the fire blanket, the understanding of the occupant on how to use it, and the response time in applying the mitigation action. Should the blanket not have been present, the occupant not understanding how to use it, or the response time of the occupant not being fast enough, this mitigation action may be unsuccessful. Also, if the detection element is too sensitive that it may have been a regular cause for false alarm, which may impact on the mitigation effectiveness. For example, the occupant's response time may be delayed on the basis of their assumption that the alarm is a false alarm. Consider also the occupant is not present. The only response is therefore an audible alarm which does nothing to mitigate the fire.

Any of these factors of mitigation effectiveness would mean the reasonably small pot fire which is easily controlled, becomes uninhibited in its growth. This could potentially result in loss of the entire building.

This simple example shows the importance of the mitigation action. It also demonstrates the importance of several other factors, for example, the adequacy of the mitigation measures, and the criticality of their presence and effectiveness. It also shows the importance of intervention of the mitigation action when the fire event is still easily controllable.

This demonstrates the philosophy of mitigation with respect to fire and gas detection: detect small to prevent uncontrollable escalation. The targets for F&G detection must allow for the fire or gas release to be detected at an early enough stage whereby we give the mitigation action the best possible chance to inhibit the escalation and prevent a reasonably small event becoming unmanageable. Equally, it shows that we should not be designing to detect events so small that false alarms, or alarms to events which present little or no danger, can occur. Such sensitivity can lead to a systemic cause for mitigation failure. An example of this in the process industry would be facility shutdown on the detection of every leak of gas, regardless of volume and volatility of the gas.

In the context of the hazardous industries, mitigation actions can vary. The mitigation action could be as simple as a general alarm to evacuate personnel, to a full facility shutdown of hydrocarbon processing, with isolation of electrical/ignition sources.

The extent of the mitigation action should be proportional to and effective against the risk being managed. It is important that the F&G detection is designed with this in mind. Certain aspects of the mitigation action may even be driven by the specific functionality of the detection system. Examples include any manual response to the incident being influenced by the detection (for example, detection of a fire by CCTV-based flame detection can allow more effective manual mitigation measures); the effectiveness of the response time of the detection (for example, an ultra-fast response flame detection device may be required where the mitigation is required immediately); the pin pointing of the location of the incident to allow effective actions specifically in that particular location (for example, voting of detectors in a fire zone to allow local confirmed detection of a fire, allowing a pinpointed shutdown action for that fire zone, while allowing other areas to continue their operation).

The variance in mitigation action can be substantial and must be considered in the context of a performance-based design to ensure adequacy to the specific risks. There is a continual relationship between detection and mitigation action which should always be considered. A useful source on mitigation is present in the form of the UKOOA Fire and Explosion Guidance (13).

The effectiveness of the mitigation is heavily dependent on the effectiveness of the detection. The detection, therefore, has to be effective with the intention of the mitigation action in mind in order to be successful in diverting the path towards a major event. This makes the setting of detection performance targets critical.

One example where the unique nature of mitigation action coupled with detection can be challenging is the application of flammable gas detection in external processing areas. The issue (with both mitigation and detection design) is that gas is a moving target.

Consider a mitigation action in Area A whereby upon detection of flammable gas, the flow of hydrocarbons is shut down and the flow is diverted elsewhere. Consider now that the gas which was detected has travelled to that area from a pressurised pipe rupture in Area B, adjacent to Area A. The momentum of the release could push the gas to any location across the facility, and on this particular day, the release orientation points towards Area A. As Area A presents an explosion hazard itself due to flammable gas processing in the area, it has its own flammable gas detection through the volume. The released gas from Area B travels to Area A and provides gas alarms from the detection. The mitigation action shuts down the hydrocarbons in Area A, diverting them to Area B . . . where the leak is present. This demonstrates an inadequate, and dangerous, mitigation action.

The purpose of this book is not to address specific mitigation measures, but rather to highlight the importance of considering such factors in the placement of detection, and the subsequent mitigation action, in the context of the anticipated hazards. Only then can a truly performance-based system be implemented to ensure that relatively minor incidents are inhibited from progressing to major accidents. Within the UKOOA Fire and Explosion Guidance, the notion of mitigation variance is discussed. Three bands of mitigation are presented from A to C, increasing in complexity and novelty along the chain. While the preference is moving a design towards A (being the most straightforward design principle), the document accepts the nature of risk and allows for performance-based assessment of the safety system and function. This is critical in F&G detection design and the associated mitigation actions.

In light of this, it is also important to note that the best mitigation is to prevent an event in the first place, but as this type of 'mitigation' falls under 'prevention', this will not be fully expanded in this book. When mitigation is discussed in this book, it will be in the context of post incident/event (i.e. after a fire or gas release has occurred and the design intention is to prevent this from growing into a major or unmanageable accident). More on the nature of systemic safety is presented in Chapter 3 which addresses the nature of systemic risk which will naturally address the issues associated with prevention and mitigation of incidents and accidents.

The relationship between prevention and mitigation, and the route towards a major accident hazard, is perhaps best simplified in the UKOOA's guidance that control is better than mitigation, which is better than emergency response.

References

1. Qureshi ZH. A review of accident modelling approaches for complex socio-technical systems. In: Proceedings of the Twelfth Australian Workshop on Safety Critical Systems and Software and Safety-related Programmable Systems—Volume 86. Adelaide: Australian Computer Society, Inc.; 2007. p. 47–59.
2. Leveson N. A new accident model for engineering safer systems. Safety Science. 2004;42(4):237–70.
3. Leveson N. Engineering a Safer World: Systems Thinking Applied to Safety. The MIT Press; 2012.
4. McNay J, Puisa R, Vassalos D. Analysis of effectiveness of fire safety in machinery spaces. Fire Safety Journal. 2019;108:102859.
5. Wikman J, Evegren F, Rahm M, Leroux J, Breuillard A, Kjellberg M, et al. Study Investigating Cost Effective Measures for Reducing the Risk from Fires on Ro-Ro Passenger Ships (FIRESAFE). European Maritime Safety Agency; 2017.
6. Rae A, McDermid J, Alexander R. The science and superstition of quantitative risk assessment. Journal of Systems Safety. 2012;48(4):28.
7. Beer J. The True Significance of Common Cause Failures in Accidents. University of York; 2011.
8. Machol R. Principles of operations research—10. The titanic coincidence. Interfaces. 1975;5(3):53–4.
9. BSI. BS5839–6:2019 Fire Detection and Fire Alarm Systems for Buildings. Code of Practice for the Design, Installation, Commissioning and Maintenance of Fire Detection and Fire Alarm Systems in Domestic Premises. BSI Standards Limited; 2019.
10. Her Majesty's Government. The Building Regulations 2010 Approved Document B. Department for Levelling Up, Housing and Communities; 2019.
11. The Scottish Government. Building Standards Technical Handbook 2020: Non-domestic. Local Government and Communities Directorate; 2020.
12. BSI. BS7974: 2019 Application of Fire Safety Engineering Principles to the Design of Buildings—Code of Practice. BSI Standards Ltd.; 2019.
13. UKOOA. Fire and Explosion Guidance Part 1: Avoidance and Mitigation of Explosions Issue 1. UK Offshore Operators Association; 2003.

3 Systemic Safety

Fire is widely accepted to pose a significant risk in the hazardous industries, with the societal shift towards safer operations being singled out by operators as a primary focus. We do, however, continue to see fires occur in these facilities across the globe.

While the content of this book will focus on the practicalities of F&G detection design, it is critical to be cognisant of the systemic nature of achieving fire and explosion safety. Adequate and proportional detection design is important, but it is one small part of ensuring safety against the risk of fire and explosion. While traditional approaches to the design of safety systems are prevalent in the industry (as will be discussed in this chapter), there may be an opportunity to strengthen fire and explosion safety through alternative emerging systemic methods.

If one adopts the notion of hazard control, fires and breaks in containment continually occur, because the control is inadequate at certain critical moments. These moments may occur at the sharp end within the process area, or they may occur far from that location, for example, through managerial decision making in the offices of the facility owner. To control a hazard, there is a hierarchical structure in place, from engineers and equipment, to the company (who design, update, and enforce the safety case), and beyond to regulators and designers (who verify and impose constraints on the operations and accepted conditions). Ultimately, therefore, there must be something wrong in this system if it cannot achieve its purpose of fire and explosion free operations.

Within the process industry, it is credible that a novel structured systemic approach, which accounts for goal-based fire safety, can be used to demonstrate alternative design arrangements of equivalent safety from traditional reactionary approaches. This will assist fire and gas safety measures in more adequately encompassing the preventive region (rather than including generic likelihood of component failure in an event tree, which subsequently influences detection measures, for example), moving towards a system-based, holistic analysis. This approach will help build resilience, where the design of one safety measure is not determined by the likelihood of another failing.

DOI: 10.1201/9781003246725-3

Resilience in this context refers to 'an organisation's ability to detect, prevent, respond to, and recover and learn from operational and technological failures which may impact the delivery of critical business and economic functions or underlying business services' (1).

Systemic hazard analysis methodologies allow the application of systemic hazard analysis techniques higher in the safety control hierarchy (e.g. the regulators, operating organisation management) in addition to the sharp end (e.g. field equipment) (2), which opens up an interesting area of further investigation, to determine how best to incorporate this methodology in the improvement of individual safety component design.

Ultimately, this approach can be applied virtually within any field which requires hazard prevention, where a control structure is present encompassing humans, technology, and the environment (the socio-technical system), with hazard prevention measures implemented in design and operation. Such fields include, but are not limited to, the aviation, business, and finance sectors. The benefit of the approach is presented in demonstrating that the combination of a focus on prevention, a structured decomposition of the system of control, and the direct connection between 'design' and 'operations' can result in any 'event' becoming preventable through adequate systemic control. One such example is the application of systemic approaches by governments in addressing societal issues such as violent crime (3). When the prevention of a problem such as knife crime is treated as an issue of systemic control, rather than the mitigation through law and order, opportunities to address previously unrealised causal factors emerge, showing a wide application potential of the approach.

This may seem odd in the context of F&G design, but it is important to note the difference in approach with respect to prevention. When safety analysis deals with prevention and mitigation in the same way, i.e. by assigning a probability of failure in an event tree, modern safety engineering suggests that this may not be optimal. As will be discussed, the design of F&G detection may be best analysed exclusively of consideration of the likelihood of the event occurring. Rather, it may be more pertinent to assume that an incident has occurred and ensure that the system is adequate from there. The systemic approach linking all the facets of fire prevention and mitigation can then be analysed with the knowledge that each component has been designed appropriately, without reliance on the likelihood of one event occurring in a particular way.

It is often suggested that the prime focus of oil and gas safety falls on accident mitigation, as opposed to prevention (4). Anecdotally, this belief is frequently explained by referring to supposedly higher cost-effectiveness of accident mitigation measures. This implies that, somewhat misleadingly, a rational strategy would be to start with improving mitigating measures (e.g. fire detection and suppression), and only then improve preventive barriers (e.g. detection of fire precursors). Therefore, while both prevention and mitigation are clearly present, one must analyse the overall philosophy of fire safety and how these facets of fire safety interact to provide safe operation.

The significance of the proposition that mitigation is more cost-effective goes beyond the accurate understanding of the state-of-the-art on fire safety. If it turns out to be wrong, it can misplace precious resources aimed at safety improvements. While this book will not investigate the accuracy of this assumption, it is important to also highlight the importance of prevention, and the importance of the design of systemic safety.

Risk Assessment Techniques

The two prevalent traditional accident models are sequential [i.e. domino effect (5)] and epidemiological [i.e. Swiss cheese (6)]. Sequential models would treat the problem of incident occurrence as one failure leading to another, leading to another, and so on, thus the analogy with the domino effect. Epidemiological models, however, would incorporate latent failures into the model. Epidemiological models maintain a similar 'trajectory' as sequential models, in the sense that an arrow of direction and linearity remains present and fundamental in their application.

The epidemiological approach first put forward by Gordon in the 1940s suggests that accidents are caused by random interactions between the agent (energy), the environment, and the host (victim). While this was intended to look at disease, it was determined as pertinent in looking at accidents also. This, as with most other hazard analysis and identification techniques, applied a reactionary approach to risk.

Qureshi (7) concludes that neither of these models is adequate in the analysis of complex modern systems.

With respect to the design and review of safety systems, multiple risk/accident models have historically been applied, with those previously discussed traditional models remaining prevalent in the process industries. More recently, systemic-based models have risen in prevalence in adjacent industries such as military and aerospace applications (7, 8), but are yet to be applied in mainstream oil and gas risk analysis. This is despite such an approach, using STPA, for example, showing some evidence of higher effectiveness in reducing likelihood of an incident occurring (9).

While the application of systemic risk analysis may not seem relevant to designing F&G detection systems, the fact that traditional approaches (such as event trees) are being used to justify detection coverage shows the creep of cost-effectiveness analysis of F&G devices being influenced by adjacent 'layers' of protection. This makes systemic analysis crucial from a high-level review of fire and explosion safety. The application of such traditional techniques, accounting for adjacent safety measures and event likelihood values to justify detection design, may in itself be a systemic route to failure.

When reviewing the cause-and-effect-based analysis applied in traditional hazard analysis, evidence shows that these analyses typically start from commonly known failures (10), as seen in the marine industry with failures such as a release of oil mist contacting a hot surface. Goerlandt and Montewka

(11) state that calculations of probabilities typically come from observed frequencies. As these traditional models do not address the entire socio-technical context, dealing with the interactions between humans, technology, and the organisation, it is challenging for them to address lagging failures interacting with one another (12) which may occur further left on the bow-tie, or higher up in the control hierarchy.

Qureshi (7) critiques historical risk assessment methods in not keeping up to speed with technological advances. The fact that causes of modern disasters vary from historical events means that in order to prevent future events, a new method of risk assessment, such as the systemic technique proposed by Leveson (13), is required.

In reviewing the limitations of the traditional approach to safety, Schröder-Hinrichs et al. (14) discuss the analysis of 41 accident investigations relating to fires/explosions in machinery enclosures. The results show that considering the expectations of the IMO guidelines with respect to accident investigations, organisational factors were not adequately addressed. Their findings show that a large percentage of the stated accident causes related to technological failures and failed physical safety barriers. This critique highlights the limitations in looking at proximate events and fire propagation, and notes that failures further left in the bowtie may not be being highlighted using the current approach. The findings are further examined by Rollenhagen (15), who demonstrates that there is a lack of attention to the organisational context in which an accident investigation is taking place.

Existing methods are informed by experience, fire safety tradition, technological and programmatic constraints, as well as the input from accident investigation. The latter plays a special role, for it arguably creates new knowledge by explaining what happened and how this can be avoided next time (16).

Accident Investigation

Accident investigations begin with prior assumptions at hand, i.e. an accident model—the understanding of how fires, and accidents in general, occur. The prior assumptions prompt what to look for and what will, therefore, be found (17). Subsequently, what is found is recommended to be fixed. The problem is that these assumptions can be flawed on some fundamental aspects, leading to ineffective safety rules, regulations, good practices, and, in part, safety systems. This can be seen in the analysis of undetected gas releases in the North Sea, automatically being attributed to ineffective gas detection layouts. As there is a tendency in the industry to critique the traditional methods of gas detection placement when undetected gas releases are reported, the analysts find what they look for rather than scientifically addressing the evidence at hand. This is addressed later in this book by Hilditch et al. (18).

This results in inadequate hazard control and, therefore, incidents. On the other hand, ineffectiveness may also be unwittingly introduced by political, commercial, and other pressures. Statutory safety rules, in particular, are a result of a consensus between rule givers (i.e. regulators) and rules takers (i.e. operators and designers) (19). Thus, in the strict sense, safety rules cannot be said to set either a minimal or average safety level, but merely an agreed safety level.

Flaws in prior assumptions about how incidents occur, and hence how they can be prevented, have attracted much attention over the last few decades. Pre-ignition events take time to develop. Conditions (e.g. incorrect design assumptions and presence of design limitations) would lead to events or other conditions (e.g. ill-informed management, training, and O&M procedures), which in turn lead to other events and conditions and so on. The metaphors like 'incubation period' and 'drifting into failure' are used to explain the dormant, latent conditions in a system that, with time, insidiously degrade the system to the point when an incident becomes imminent (20, 21). This degradation, or drift, is systematic (not random) and fuelled by the natural phenomena of adaptation to new circumstances (endogenous and exogenous) and optimisation of resources (22); it is also explained as the inexorable manifestation of entropy. A helpful property of this dynamic is its relative slowness and determinism, which means that the drift is detectable and preventable (23, 24). On the other hand, this natural system dynamic undermines the utility of accident investigation results that inevitably reflect the past circumstances, which may not remain relevant (25). It also shows the importance of maintaining relevance of safety measures and barriers throughout the lifecycle of a facility. Assumptions during design of F&G detection systems regarding incident likelihood may, therefore, be somewhat ill advised.

When reviewing accident causation, it is important to consider the complexity in the context, and factors which lead to an incident. For example, there may be eight causes of a hazard, and when all eight causes are present together, the hazard is realised. It may be the case, however, that causes 1 and 3 together will result in the same hazard as causes 2, 4, and 6 together. How can a hazard identification and analysis approach consider such combinations for a complex system accounting for the interactions between humans, technology, and the environment?

Consider a systemic causal factor where incident A may result in B, but incident B does not require A to occur. Vehicle accidents as a result of drink driving are a prime example of this. Driving while under the influence of alcohol does not always result in an accident, but it is a potential cause (26). It is therefore important that causal factors are considered for all unsafe control actions which may be taken. For a control action to be unsafe, the context must be relevant for that action to become unsafe. By understanding these contexts and actions, causal scenarios can be identified and eliminated.

Cost-Effectiveness

While an endless number of potential scenarios can be developed and expanded upon, and necessary controls generated, one must consider that resources are finite. Cost-effectiveness is a critical facet in the design of any safety system. Ultimately, resources are finite and difficult decisions need to be made when deciding where investment is to be placed. Philosophically, however, can we negate either a preventive or mitigation safety barrier due to the perceived strength of the other?

The reference to cost-effectiveness usually refers to the criteria applied during a QRA. There can, however, also be subjective or experiential judgement leading to this belief. It can, for instance, be stimulated by the perception that the complexity of modern designs and operations makes incidents inevitable. Indeed, in complex system design, errors are frequent and procedures are often underspecified (27). Hence, improving the response through higher robustness of passive systems and resilience of active ones should take precedence over, arguably, contingent effort on prevention. For instance, fire protection in machinery spaces has recently been transformed by high-pressure water mist systems (28–30) and new generations of fire alarm systems that use intelligent sensors (31).

Regardless of how much work is done on accident mitigation, it only underlines the importance of this second barrier being dependent upon failure of the first barrier: prevention. Importantly, from the cost-effectiveness perspective, the main effort may actually need to be directed to prevention. It is commonly known that preventing a hazard from materialising is more preferable than fighting its consequences (32–34). If prevention succeeds, the event is only a near-miss, (or potentially not even registered as a near miss if the drift towards failure is detected and prevented early enough), thereby normal operations will not be disturbed. Eliminating the hazard from the system has to be the best strategy, provided the facility's prime function and its performance requirements remain undisturbed. By eliminating a hazard, a myriad of opportunities for various incidents are singularly eliminated. If a hazard cannot be eliminated, minimising the likelihood of development into specific incidents by some control actions would still be more effective than the alleviation of incident damages. Additionally, the cost of implementing preventive strategies has been shown to be generally lower than the cost of reliance on mitigation barriers (9, 23, 35). Even with the greatest effort on prevention, however, fire events can still occur.

When reviewing hazardous areas in the marine industry, in the top causes of loss where this resulted in a claim, fire ranked second on the list by the number of claims at 16%, and also second in value of claim (36). Between August 2011 and January 2018, among 112 'serious' and 'very serious' fire incidents (as defined by EMSA), 57 were in the machinery/ hazardous spaces (37). More than a decade ago, a study by Det Norske Veritas (DNV) reported that about half of fires that occurred in machinery spaces resulted from the contact between combustible oil mist and high temperature surfaces (38). Little has changed since then, for this very scenario remains rife today (39).

The approach to fire safety is intended to reflect knowledge of how fires happen and develop, and, therefore, implement methods whereby fires shall be prevented or at least controlled. Prevention and control (or mitigation) are often considered complementary and could be viewed as two sequential processes of protection, providing redundancy in safety against fire development or break in containment. In fact, idealistically these could be classed as independent. If the design can, for example, eliminate the flammable material entirely, there could be no need for a fire detection and suppression system. While this is idealistic, it is critical in the argument for cost-effectiveness when we consider that prevention and mitigation are closely related and trade-offs can therefore be struck.

Does Defence in Depth Work?

The important realisation is that the underlying system behaviour cannot be changed by merely looking for or reacting to its events, or by analysing components of the system in isolation (40). Therefore, increasing the layers of protection does not necessarily lead to safety in socio-technical systems, because additional safety barriers and protection can be defeated by psychological reactions (41), and designing based on the likelihood of such barrier failure is missing the point on what causes accidents.

The reason is that a socio-technical system is more than the sum of its components (22). Thus, the cause of an incident and an accident is the inadequate design of the system as a whole, rather than specific scenarios in isolation. Such inadequacies, e.g. flawed links, can exist for a brief moment in time in the correct context, yet distant to each other in a different time and context. This can, however, be enough to cause an effect that is nonlinear, such as small events or conditions which can cause serious consequences (42); the reverse is also true. Therefore, the modern paradigm of system thinking may be more appropriate. The systems' approach accentuates the importance of nonlinear interactions between system components and the system structure and mental models that determine it (43).

This paradigm is taken up by safety research. It is observed that such research points towards further appreciation of the organisation, inter-organisational, and human aspect of the problem (7, 44–46). Such research also points to the importance of not analysing barriers in isolation, but rather applying an appreciation of the entire system, and the interactions between its components, being of critical importance when analysing the emergence of 'safety' as a condition. While this wider safety research does not directly relate to fire and gas safety in the hazardous industries, its tailored application to fire and explosion safety at large in the hazardous industry would be novel and may provide holistic improvements. This systemic approach to accident prevention should not be at the expense, however, of adequate design of individual facets of fire and explosion safety.

With the aforementioned consideration in mind, the question is whether the prior assumptions that guide accident analysis, safety rules, and regulations, etc. are indeed flawed. If so, we should not expect effective prevention or/and mitigation of fire events. In fact, as systems include an increased human, environment, and technological interaction, should we expect fire incidents to increase if the approach to fire and explosion safety does not also evolve and continues to simply apply the traditional event tree analysis type of QRA process? This is an important point when we consider adding complexity into the F&G design—is this counter intuitive to successfully achieving a safe state?

Systemic Influence in F&G Design

Taking a closer look at incident prevention, the bias towards direct causal factors may be attributed to the widespread application of these traditional linear, event-based models to accident analysis, namely, the previously discussed sequential and epidemiological models [the Domino (5) and Swiss cheese (6) metaphors]. These models represent the classic paradigm of linear representation of causation, where clear links between causes and effects must be known, and system safety is assumingly improved by independent treatment of individual components of the system. Hence, the system is assumed as essentially a sum of its individual components. Although this linear paradigm has its own merits, it is insufficient to achieve system safety as shown in the literature (7–9, 12, 44, 47–49). The linear thinking focuses on incident events and their patterns—which are the visible tip of the iceberg—but systematically fails to explain the underlying structure and mental models that give rise to those events—the invisible, underlying portion of the iceberg (50). The underlying system behaviour cannot be changed by merely reacting to its events, or by analysing components of the system in isolation (40). The reason is that a system is more than the sum of its components (22). Therefore, if we are to apply a holistic approach to safety when designing the F&G system, the modern paradigm of system thinking could be addressed. In addition to its contribution to causation of individual components, the systems' approach accentuates the importance of nonlinear interactions between them and the system structure and mental models that determine it (43). As a result, the accident analysis will naturally seek to answer not only what happened but also why it happened. This finding presents a significant gap worth investigating that a more structured approach to fire safety, accounting for the entire system specifically further left on the bowtie, will present improvements, but may not be best placed in any quantitative analysis of individual barrier performance.

Such techniques do exist, including the SCAT chart—Systematic Cause Analysis Technique (51) and SHORE (2). These accident/incident analysis and systemic design and operational techniques attempt to address direct causes, as well as the basic/underlying causes of events.

While mitigation begins from the assumption of a fire being present (likelihood of fire = 1.0), preventive categories do not work in this way. The traditional approach of assigning event consequence, as we do with mitigation systems (i.e. if sprinklers do not activate, what RHO is generated from the fire), may not be an accurate method of analysis when compared to an analysis of the entire system effectiveness (i.e. the ability of the system to account for all preventive categories rather than each in isolation).

This further demonstrates that in order to design and operate an effective system, a different approach may have to be applied. Mitigation, where cost–benefit analysis can more easily be carried out, along with technological capabilities which allow mitigation barrier health to be monitored, may continue to apply a quantitative risk approach in deciding what level of coverage to apply. When it comes to prevention, however, such an approach may need to be supplemented with a systems-based technique which can only be effective when the entire left-hand side of the bowtie (for want of a graphical representation) is accounted for. This allows the operators to focus on the interactions between all of the actors in the system, rather than a reductionist view of safety being a sum of all the parts, leading to the criticised approach of adding more layers of barriers to 'increase' safety (40). More concerning still is supplementing the event tree with addition pre-release events (with associated likelihood value) to dilute the requirements of a mitigation barrier like F&G detection as previously discussed.

Applying a systemic approach is one in which the groundwork already exists, with most global regions allowing for performance-based design. With the ability to address alternative design arrangements, we move towards a goal-based approach which requires designs to set functional requirements allowing designers the freedom to meet those requirements in any number of ways. With the move away from prescriptive guidance, this provides the scope for the process industries to expand from a focus on typical generic proximate events, towards addressing prevention more thoroughly and effectively as a complete system.

The principle is essentially deterministic, strictly demanding to have a barrier. In this case, both preventive and mitigative barriers are required, and weakness of one barrier cannot be compensated by making another one stronger (52). The rationale is that even the strongest barrier can fail under certain foreseeable (albeit unlikely or even reasonable) scenarios. This demonstrates the importance of treating the problem from a systemic perspective, and highlights the importance of verifying that soft, as well as hard, safety factors are being implemented in operation, as assumed during design. This shows that regardless of the cost-effectiveness argument, auditing/operational processes of measurement will play a crucial role in systemic operational safety.

It is also true that prevention and mitigation essentially deal with different phenomena that require a different response. Pre-incident events take time to develop through systematic (not random) processes, with detectable warning signs and recognisable precursors (23). The metaphors like 'incubation period' and 'drifting into failure' are used to explain the insidious degradation of the system up to the point when an incident becomes imminent (20, 21). This degradation, or drift to failure, is a result of normal dynamics of systems, i.e. constant adaptation (adjustments) and optimisation of resources to daily circumstances (22, 53). Incidents happen when interactions between system components (hardware, software, people, and organisations) become dysfunctional, even though individual components have not failed (25). When such interactions are adequate, even sudden failures of individual components (e.g. power loss in stormy weather) would be effectively mitigated. Thus, incident prevention operates within a complex socio-technical system (54), and it relies on effective safety management.

In contrast, mitigation is heavily reliant on technology, i.e. engineering controls. From the physical point of view, incidents develop to accidents when the response energy is unable to thwart the energy being released. Speed of response is decisive, for the damage energy is released rapidly (e.g. a gas jet fire from a pressurised vessel). Engineering controls, particularly passive ones, have proven to be effective in such abnormal, distress conditions. The role of people is critical, but they arguably would be unable to compensate for a bad design.

There exists, however, an issue in addressing the notion of holistic fire safety, accounting for fire prevention and mitigation in the same analysis. In summary of the previously discussed issue, if prevention, for example, is included in the overall risk analysis as part of an event tree analysis, the overall risk-reduction capability of mitigation measures can be diluted. Where likelihood of initiating events further left on the bowtie are accounted for, the likelihood of the central event (i.e. fire) is of such a low likelihood, that the risk benefit of mitigation measures are reduced to the point of potential omission from a cost–benefit perspective.

Typically, in deterministic fire safety engineering, the likelihood of the central initiating event (i.e. fire) is taken as 1.0 to ensure that sufficient mitigation measures are applied. While this may overestimate the cost-effectiveness of mitigation, there is a risk in moving the other direction. Including likelihood calculations of initiating events in the mitigation cost-effectiveness calculation will ultimately result in a reduction in mitigation 'effectiveness'. This dilution of the risk-reduction measures attributed to mitigation could be dangerous in reducing the emphasis on technology-based mitigation.

References

1. Protiviti. The Road to Resiliency: Building a Robust Audit Plan for Operational Resilience. Available from: https://www.protiviti.com/UK-en/insights/whitepaper-road-resiliency.

2. McNay J. A Structured, Systemic Methodology to Improve Maritime Fire Safety in Machinery Spaces. University of Strathclyde; 2020.
3. BBC. How Scotland Stemmed the Tide of Knife Crime 2019. Available from: www.bbc.co.uk/news/uk-scotland-45572691.
4. Holmberg JE. Defense-in-depth. In: Handbook of Safety Principles. John Wiley & Sons, Inc.; 2017.
5. Heinrich H, Peterson D, Roos N. Industrial Accident Prevention. 5th ed. McGraw Hill; 1980.
6. Reason J. Human Error. Cambridge University Press; 1990.
7. Qureshi ZH. A review of accident modelling approaches for complex socio-technical systems. In: Proceedings of the Twelfth Australian Workshop on Safety Critical Systems and Software and Safety-related Programmable Systems—Volume 86. Adelaide: Australian Computer Society, Inc.; 2007. p. 47–59.
8. Everett C, Hall T, Insley S. NASA Accident Precursor Analysis Handbook. National Aeronautics and Space Administration Office of Safety and Mission Assurance; 2011.
9. Wróbel K, Montewka J, Kujala P. System-theoretic approach to safety of remotely-controlled merchant vessel. Ocean Engineering. 2018;152:334–45.
10. Pomeroy R, Earthy J. Merchant shipping's reliance on learning from incidents—A habit that needs to change for a challenging future. Safety Science Journal. 2017;99:45–57.
11. Goerlandt F, Montewka J. Maritime transportation risk analysis: Review and analysis in light of some foundational issues. Reliability Engineering & System Safety. 2015;138:115–34.
12. Carroll J. Organizational Learning Activities in High-hazard Industries: The Logics Underlying Self-analysis. John Wiley & Sons, Inc.; 1998. p. 699–717.
13. Leveson N. A new accident model for engineering safer systems. Safety Science. 2004;42(4):237–70.
14. Schröder-Hinrichs JU, Baldauf M, Ghirxi KT. Accident investigation reporting deficiencies related to organizational factors in machinery space fires and explosions. Accident Analysis & Prevention. 2011;43(3):1187–96.
15. Rollenhagen C, Westerlund J, Lundberg J, Hollnagel E. The context and habits of accident investigation practices: A study of 108 Swedish investigators. Safety Science. 2010;48(7):859–67.
16. McNay J, Puisa R, Vassalos D. Analysis of effectiveness of fire safety in machinery spaces. Fire Safety Journal. 2019;108.
17. Lundberg J, Rollenhagen C, Hollnagel E. What-you-look-for-is-what-you-find—The consequences of underlying accident models in eight accident investigation manuals. Safety Science. 2009;47(10):1297–311.
18. Hilditch R, McNay J. Addressing the problem of poor gas leak detection rates on UK offshore platforms. In: Proceedings of the Ninth International Seminar on Fire and Explosion Hazards. Vol. 2: 21–26 April 2019. Saint Petersburg, Russia; 2019.
19. Kristiansen S. Maritime Transportation: Safety Management and Risk Analysis. 1st ed. Routledge; 2005.
20. Turner BA. Disasters, Man-Made. Wykeham Publications; 1978.
21. Dekker S, Pruchnicki S. Drifting into failure: theorising the dynamics of disaster incubation. Theoretical Issues in Ergonomics Science. 2014;15(6):534–44.
22. Rasmussen J. Risk management in a dynamic society: a modelling problem. Safety Science. 1997;27(2):183–213.

23. Leveson N. Engineering a Safer World: Systems Thinking Applied to Safety. The MIT Press; 2012.
24. Rasmussen J. Risk Management, Adaptation, and Design for Safety. Springer; 1994.
25. Leveson NG. Applying systems thinking to analyze and learn from events. Safety Science. 2011;49(1):55–64.
26. Leveson N. An STPA Primer. Available from: https://fliphtml5.com/sgqs/syzv.
27. Perrow C. Normal Accidents: Living with High Risk Technologies-Updated Edition. Princeton University Press; 2011.
28. Kääriäinen JS, editor. High Pressure Water Mist Fire Protection Systems. ASME Turbo Expo 2007: Power for Land, Sea, and Air. American Society of Mechanical Engineers; 2007.
29. Liu Z, Kim AK. A review of water mist fire suppression systems—fundamental studies. Journal of Fire Protection Engineering. 1999;10(3):32–50.
30. Arvidson M. Large-scale water spray and water mist fire suppression system tests for the protection of Ro—Ro cargo decks on ships. Fire Technology. 2014; 50(3):589–610.
31. Bistrović M, Kezić D, Komorčec D. Historical Development of Fire Detection System Technology on Ships. Available from: https://hrcak.srce.hr/1126212013.
32. Möller N, Hansson SO. Principles of engineering safety: Risk and uncertainty reduction. Reliability Engineering & System Safety. 2008;93(6):798–805.
33. Bahr NJ. System Safety Engineering and Risk Assessment: A Practical Approach. CRC Press; 2014.
34. Hollnagel E. Barriers and Accident Prevention. Routledge; 2016.
35. Lees F, Mannan S. Lees' Loss Prevention in the Process Industries, Hazard Identification, Assessment and Control. 4th ed. Butterworth-Heinemann; 2012.
36. Benito L, editor. Taken from Big Data Technology for Maritime Safety. IMO International Conference. Busan, South Korea; 2017.
37. EMSA. Published Maritime Casualty Investigation Reports 2018. Available from: https://emcipportal.jrc.ec.europa.eu/index.php?id=44.
38. Engine Room Fires Can Be Avoided [press release]. Der Norske Veritas; 2000.
39. DNVGL. Recommended Practice: Engine Room Fire Prevention. Report No.: DNVGL-RP-0279; 2018 January.
40. Read G, Salmon P, Lenné M. Sounding the warning bells: The need for a systems approach to understanding behaviour at rail level crossings. Applied Ergonomics. 2013;44(5):764–74.
41. Besnard D, Hollnagel E. I want to believe: Some myths about the management of industrial safety. Cognition, Technology & Work. 2014;16:13–23.
42. Hollnagel E. Safety-I and Safety-II: The Past and Future of Safety Management. CRC Press; 2018.
43. Meadows DH. Thinking in Systems: A Primer. Chelsea Green Publishing; 2008.
44. Leveson N, Dulac N. Incorporating safety in early system architecture trade studies. Journal of Spacecraft and Rockets. 2009;46(2):430–7.
45. Dekker S. Drift into Failure: From Hunting Broken Components to Understanding Complex Systems. Ashgate; 2011.
46. Rokseth B, Utne IB, Vinnem JE. A systems approach to risk analysis of maritime operations. Proceedings of the Institution of Mechanical Engineers, Part O: Journal of Risk and Reliability. 2017;231(1):53–68.
47. Lundberg J, Rollenhagen C, Hollnagel E. What you find is not always what you fix—How other aspects than causes of accidents decide recommendations for remedial actions. Accident Analysis & Prevention. 2010;42(6):2132–9.

48. Hollnagel E. Risk+barriers=safety? Safety Science. 2008;46(2):221–9.
49. Carroll J. Incident reviews in high-hazard industries: Sensemaking and learning under ambiguity and accountability. Industrial and Environmental Crisis Quarterly. 1995;9(2):175–97.
50. Kim DH. Introduction to Systems Thinking. Pegasus Communications; 1999.
51. DNVGL. Incident Investigation: Expert Analysis Is the Key to Preventing Recurrences. Available from: www.dnvgl.com/services/incident-investigation-1095.
52. Holmberg JE. Defense-in-depth. In: Handbook of Safety Principles. John Wiley & Sons, Inc.; 2017. p. 42–62.
53. Hollnagel E. The ETTO Principle: Efficiency-thoroughness Trade-off: Why Things That Go Right Sometimes Go Wrong. CRC Press; 2017.
54. Puisa R, Lin L, Bolbot V, Vassalos D. Unravelling causal factors of maritime incidents and accidents. Safety Science. 2018;110:124–41.

4 Flame Detection Technologies

For those involved in flame detection, there will be familiarity to the claims of false alarm immunity with ultra-fast response detection to very small fires at great distances. Those with operational experience will know that it is not possible at the time of writing to achieve all at the same time. Unfortunately, such claims can lead to confusion on the part of designers and facility operators. One such scenario is presented in the following.

A flame detector is experiencing false alarms. Fortunately, the detector is a visual flame detector so the manufacturer can easily retrieve the video recording of what was sending the detector into alarm.

On review of the video footage, it becomes clear to the manufacturer what is causing the alarm. The detector is looking directly at a process relief flare (in other words, a real fire) on an adjacent facility. At any point in time when the adjacent facility is flaring, the flame becomes large enough to set the detector into alarm (as a result of the inverse square law impact on detection distance, which will be discussed later in the book).

The problem has a reasonably easy fix . . . relocate or alter the orientation of the detector to remove the 'friendly fire' from the detector's field of view. The response from the site is . . . alarming.

'We know the detector is looking at that flare, can you update your detection algorithm so it doesn't false alarm to the fire anymore so we don't have to relocate it?' Perhaps a flame detector manufacturer being asked to change its device so it doesn't detect fires anymore may be a client request too far. Or would it?

When one considers this further, perhaps it would not be so surprising. When considering that the traditional infrared detectors come with sensitivity settings specifically to allow sites to reduce the capability of the detector to detect radiation, sites have experience of altering detection capability in an effort to reduce false alarm occurrence. One wonders how often a false alarm from a flame detector is followed by the sensitivity being reduced to prevent future alarms. When this occurs, I also ponder how often it is fed back into the design to make sure that the detection capability is maintained in serving its designed safety function.

DOI: 10.1201/9781003246725-4

On this design factor, another point can be taken from the story. Why was the detector looking at a flare? During design, 3D models and facility drawings rarely show the adjacent facilities. Suitable design will ensure that the devices are not looking towards flares, but traditionally in design, engineers often don't know what's next door. This should always be considered in design, particularly with the advances in access to satellite imagery. Even if it is missed during the design on paper, this should certainly be highlighted during commissioning in reviewing the device's field of view. Such factors will be reviewed further in Chapter 5.

Detecting Flaming Fires

There have been numerous flame detection technologies applied in the hazardous industries over the last few decades. Specifically within the Oil and Gas Industry, the development and implementation of flame detectors have been for the purpose of preventing future disasters like the 1988 Piper Alpha disaster. Following Piper Alpha, the drive towards improved safety in the industry assisted in the development of improved and more reliable flame detection devices and systems.

With the drive for facility operators to protect people, the facility, and the environment today, the flame detection industry is well established and continuing to innovate. The operating environment for such devices, from the Arctic to the Sahara, can be a challenge, and the performance of the device is paramount in providing the successful mitigating action.

Technology selection is therefore of great importance with consideration to the environment, as well as the anticipated fire the device is intended to detect. While some technology is best suited to detect clear burning chemical type fires like methanol, others are best suited to the detection of fires with a higher carbon content, up to diesel type fires which can produce thick dark smoke. This technology suitability to the hazard must also be coupled with the anticipated performance within the specific environment the hazard is situated within.

The age old comment that *there is no such thing as the perfect flame detector, each technology has its strengths and limitations* is true to this day. Technology selection for the specific application in question is critical, and any limitations or desensitisation of the selected technology for the specific facility must be addressed during the design.

Each of the flame detection technologies discussed in this book aims to detect the stimuli of a flame within a specific range of the electromagnetic spectrum. For further reading, one of the most in depth studies into the practicalities associated with optical flame detector placement is the work by Gottuk et al. (1), which is referenced in multiple locations within this book.

Figures 4.1 [derived from (2)] and 4.2 represent the burning characteristics of fictional fuels across the spectrum as a generic representation of how fire emissions have varying intensity across the spectrum. The characteristics of the burning fuel across the electromagnetic spectrum contrasted with the performance of the flame detector in 'monitoring' that region of the

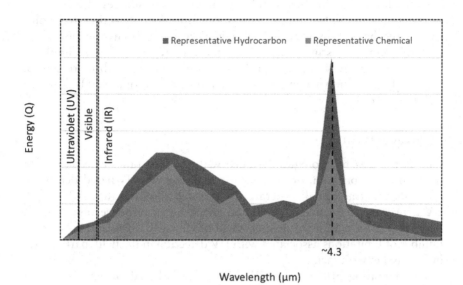

Figure 4.1 Example energy across the electromagnetic spectrum, derived from (2).

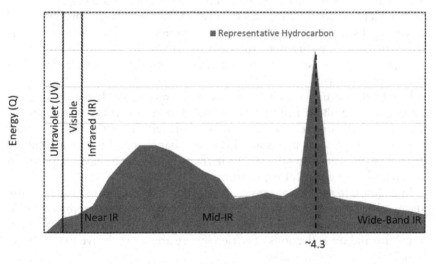

Figure 4.2 Simplified and annotated fire energy emission (Q) across the electromagnetic spectrum.

electromagnetic spectrum is what determines the performance capability of the device. This, coupled with the specific location's environmental stimuli in the region of the spectrum the detector operates, will dictate the 'in field' performance of the device.

Examples for specific fuels can be found in various sources (2).

The following sub-sections review the prevalent detection technologies in the hazardous industries today, along with discussion points on the environmental stimuli which are most relevant to that technology. Ultraviolet-, infrared-, and visual-based flame detectors will be reviewed, along with different variants of each technology (e.g. combined UV/IR and multi-channel IR).

Ultraviolet (UV)

One of the first technologies designed with the intention of detecting fire through the application of an optical-based device is UV-based detection. These devices traditionally operate at wavelengths shorter than 0.3 µm (3). UV detection is effective in providing an ultra-fast response time when a flame is present within the detector field of view (3). There are, however, a number of limitations associated with UV detection in the field, particularly in external environments.

The operating principle of UV technology is based closely to that of a Geiger Muller tube. When a UV source (e.g. a fire) is present within the field of view, UV photons striking the on board tube cathode will release an electron. This released electron subsequently strikes a gas molecule which creates the cascade effect which determines response time and detection distance of the detector. The stronger the UV source (i.e. the larger or closer the fire is to the device), the higher the count of pulses. This produces a pulse count in counts per second (CPS) which is used to determine the alarm threshold.

Due to the nature of operation in the UV region, UV-based devices are subject to a number of limitations. Degradation of performance from the presence of oil, smoke, and other obscurants, either on the lens or in the environment, presents an issue. This performance inhibition as a result of dirty environments generally results in the technology's declined application in external petrochemical applications.

In addition to the loss of sensitivity, UV flame detectors are prone to false alarm where flare radiation (either direct or reflected) can produce a sufficient UV stimulus to result in alarm. This inability to differentiate between a real fire and the reflections of a friendly fire are not exclusive to UV detection, however.

Specifically with UV detection, there is a potential for false alarm from external sources such as lightening, sparks, or arc welding (1). Such stimuli can have an adverse effect on performance with respect to reliability of detection and the potential for false alarm when placed in external or dirty environments.

It is also the case, when discussing performance, that UV detectors generally have a much shorter detection footprint when compared to more modern technologies like Triple IR and Visual detection.

The technology does, however, present some strengths which, dependent on the environment and hazard, may make it the most appropriate technology for selection.

The operating method of the device means that some of the fastest operating detectors on the market are UV based. If the environment is controlled and a strong enough source of UV is presented (i.e. from a flash fire), the response from the detector will be near instantaneous as the CPS threshold will be immediately reached and the device will immediately alarm.

In addition to this, Gottuk et al. found that the UV devices tested against other technologies performed better in the tests with obstructions within the field of view. The devices were also often the fastest detectors to respond (1). This can be of great benefit over other technologies in applications where an almost instantaneous detection response is required (for example, in some military or manufacturing applications).

Single Infrared (IR)

With the aforementioned limitations affecting UV-based flame detection, developments in flame detection technology moved on to the infrared region of the electromagnetic spectrum. Within the IR region, the devices generally focused around the 4.3 μm region. In this region, fires (particularly hydrocarbon fires) generate a spike in hot CO_2 emissions (2). This spike in hot CO_2 allows filtering in the IR region to allow devices to effectively detect, at increased ranges, specific characteristics of a flame.

This allowed technology developers to address one of the primary flaws of UV detectors. Where UV detectors would simply respond to the intensity of the UV source, IR detectors could single out specific portions of the IR region in which a flame produced a strong signal and analyse these to detect the fire. In theory, this presents the benefit of being able to reject intermittent/periodic non-fire sources of infrared radiation, while remaining responsive to genuine fires. The technology, however, remained unable to fully distinguish between direct or reflected flare radiation from 'friendly' fires. Any machinery which in normal operation would produce the products of combustion in a way which copied a real fire (e.g. combustion extracts from a turbine) also presented a false alarm source.

Additional false alarm stimuli incudes any steady-state radiator which is chopped or vibrated by the environment in such a way that the source of IR is strong enough and modulated, thus looking like a flame to the detector. Such 'fire' stimuli could be caused, for example, by mounting the detector on an unstable support and orientating the detector to view a hot object. In fact, some early models of IR technology were anecdotally known to false alarm from simply waving a warm hand in front of the lens.

In order to ensure that false alarms were not a continual occurrence, the sensitivity of these devices was relatively restricted. This presented a similar limitation to UV detection . . . a relatively short detection footprint (4), with environmental desensitisation (which will be discussed later in this book) only exacerbating this issue (5).

While the effect of desensitisation will be discussed specifically later in this book, one such factor affecting IR detection is that of water absorption. Operation within the IR region of the electromagnetic spectrum presents a challenge with respect to the transmission of IR through water in the atmosphere, or detector window. The presence of water mist or fog in the detector field of view will inhibit detection of a flame which may also be burning within the field of view of the device. The same is true of water on the faceplate of the detector, although water shields can assist in mitigating this.

With a detector operating around 4.3 μm, where said detector is mounted and looking at a flame, if water is present between the flame and the detector, the water molecules are efficient in absorbing the radiation emitted by the flame. This therefore desensitises the detector in its ability to gain a strong enough signal from the IR radiation from the flame to generate the alarm.

Specifically around 4.3 μm, water absorbs a significant portion of IR radiation, meaning that while water in the field of view affects most detection technologies, it is particularly challenging for IR-based devices to combat this (5).

Figure 4.3 [derived from (6)] represents water absorption across the electromagnetic spectrum. Note that this is simplified from other sources and has been developed for ease of reference in the flame detection discussion. The values are not to be assumed to contain a high degree of accuracy.

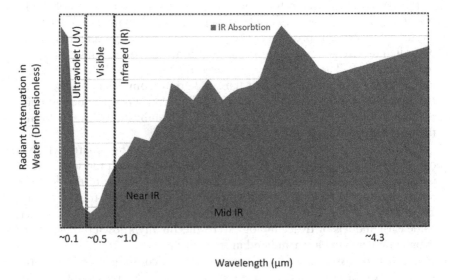

Figure 4.3 Absorption in water across the electromagnetic spectrum derived from (6).

Reference to specific sources on radiant absorption in water should be consulted for any work in this field.

As with all technologies, IR detection has some strengths which may make it the flame detector of choice for various application.

One such strength is the ability to detect fires which the human eye may not be able to detect. Fires, such as methanol, will not emit a particularly large visible flame, but will emit reasonably effectively at 4.3 µm. This makes IR-based technology useful in detecting such fires which can also include sulphur and hydrogen fires (although alteration of the filters may be required to effectively detect a hydrogen flame which does not produce a significant signal at 4.3 µm).

Performance will vary to all fires based on which filter and sensor are applied, along with how the device algorithms have been developed. This means that where an operator or designer is unsure in their interpretation of a detector manual or test report, liaising with the specific detector manufacturer to verify the detector is appropriate to the hazard and environment in question is important.

Combined UV/IR

Considering the false alarm potential from both UV- and IR-based devices, the next step in detection technology was the combined UV/IR device. The intention of the combination was to bring the benefits of each technology together and reduce false alarms. In order to achieve successful detection, both the IR and UV channels would need to register an alarm (7, 8). As there is generally no single false alarm source which would activate both 'sensors', this would seem to be suitable in theory. The technology, however, also brought forward the limitations of both technologies.

The result of this was that in the event one of the channels was inhibited in some way (for example, by sunlight, water, smoke, and radiation), detection would not occur. Considering the UV channel, this could be inhibited by dirt or oil on the window of the device, smoke in the atmosphere, etc., while the IR channel could be inhibited by background radiation or water in the field of view.

This potential 'fail to danger' in operating principle resulted in potentially unrevealed failures in external processing sites and has resulted in some operators advising against their use.

It is important despite this, however, to always address the specific hazard and environment. Depending on these two factors, the UV/IR technology may still be the most appropriate technology in meeting the design goals.

Multi-Channel IR

With the wide range of opportunity and potential associated with the IR region of the electromagnetic spectrum, technology developers began investigating improvements in the IR band by adding sensors and filters across the region. This resulted in multi-channel IR detectors, or more commonly referred, Triple IR detection.

With Triple IR detection, two (or more) reference sensors with associated filters are generally applied in order to reduce false alarm frequency associated with single-channel IR detectors. The intention was not to rely on a single channel but to take readings from three (or more) channels and analyse them in order to decide whether the source of radiation is a flame or not.

While the intention was to reduce false alarms, this had the additional benefit of allowing manufacturers to drastically increase detection distances compared to traditional single IR and UV devices (9–11).

Despite the intention of reducing false alarm occurrence, the technology (as with all flame detection technology) remains vulnerable to false alarm.

As with single IR devices, the technology remains susceptible to reflected or direct flare radiation but did see an improvement in false alarm resistance to non-flame radiative sources. As a result of this maintained susceptibility to false alarm, Triple IR detectors will still allow alteration of the sensitivity setting, often ranging from low to high sensitivity or a variation of the form.

Figure 4.4 represents the generic detection principle of Triple IR flame detection. This representation of both wavelength selection and hydrocarbon fire emissions is generic and vary, dependent on the fuel, between device manufacturer and model. For example, reference wavelengths may be either side of the measurement wavelength, or could both be on the left- or right-hand side along the x-axis.

The measurement band aims to detect the aforementioned spike in hot CO_2 at approx. 4.3 µm, while the reference bands look to analyse the elevated radiation (or lack thereof) around the measurement band. This then allows the on-board algorithms to analyse the three signals and determine if a fire is present in the field of view.

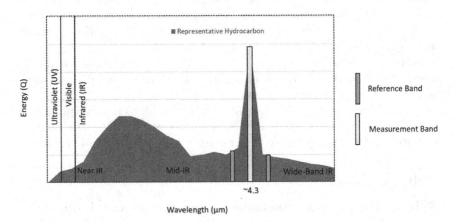

Figure 4.4 Operating principle of triple IR flame detection.

As previously mentioned, the position of the reference bands will differ based on the specific device model. This can have a significant impact on the detector's ability to detect various flames, in numerous environments, and with differing blockage configurations between the device and the flame.

While, in theory, the technology aimed to reduce false alarms to the environment as well as background radiation, the technology still presents potential for alarm in the presence of the correct stimuli. Essentially, any radiative source which can mimic a flame or the modulating products of combustion (whether visible or invisible) will present the risk of false alarms (1). As discussed, the inclusion of sensitivity settings of low, medium, and high (or a variation of those terms) (9–11) allow facilities to reduce false alarm susceptibility by reducing the sensitivity. This will subsequently, however, reduce detection capability to a fire in reality.

As with any other technology, designers must apply caution when applying the technology to a facility in which a flare is present, either on site or adjacent to the site. Consideration should also be applied when determining the sensitivity setting to be used. The continual trade-off between maximising detection distance and reducing false alarms is crucial. For example, the facility may never experience false alarms, but the detectors may only achieve a detection distance to an n-heptane square foot pan fire of 5 m.

With the move towards analysing multiple wavelengths in the IR region of the electromagnetic spectrum, the issue of desensitisation can, in theory, become more exacerbated. With single-channel IR devices, water in the FOV could desensitise the detector to a real fire. With the addition of the reference bands, any external sources which interfere with them could also present additional desensitising factors.

One such stimulus which can interfere around the 4.3 µm point so common in Triple IR devices is the sun. Sunlight interference in the 4–5 µm region can make filter selection a challenge for Triple IR manufacturers and can affect different devices in drastically different ways. In one example, this reduction in detection capability takes detection from 70 m to 9 m (an 87% reduction) (12).

As previously mentioned, this effect will vary between devices based on sensor and filter selection, but also on the programming of the on-board algorithms. Once again, this presents an example where in the absence of clarity in the detector literature, liaising with the manufacturer on their device capability is important. Manufacturers will likely keep the data on filter selection and firmware development in-house.

Figures 4.5–4.7 represent how solar transmission can interfere with the detection (or measurement) wavelength, and reference wavelengths of a Triple IR flame detector. Note that the solar transmission is representative of the drop in transmission at 4.3 µm and is not to be taken as the true value of transmission reduction.

Figure 4.5 shows the operating principle of Triple IR overlaid with a generic hydrocarbon fire. The signal of the detection central wavelength is

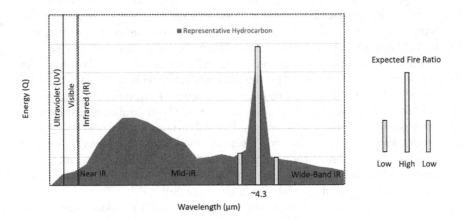

Figure 4.5 Ratio change in presence of fire.

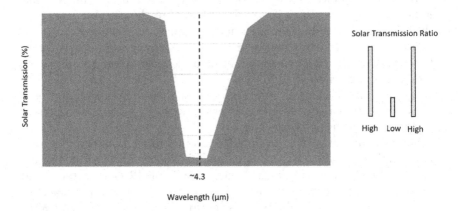

Figure 4.6 Ratio change in presence of sunlight.

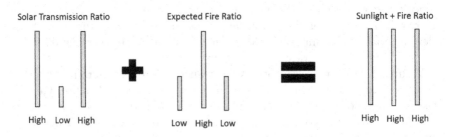

Figure 4.7 Impact of a fire + sunlight on the ratio change.

high, and the reference sensors provide a low background radiation reading. This presents a high ratio change between the detection and reference sensors, resulting in effective detection. Put in terms which may somewhat oversimplify the matter, the higher the ratio change, the greater detection distance can be achieved.

Figure 4.6 shows the impact of sunlight. This increases the readings of the reference sensors, but not the detection sensor. The result is, when a fire is present at the same time as the sun, all three sensors have an elevated reading, meaning that the ratio between detection and reference sensors is low, resulting in lower detection distance capability. This effect is often unrevealed and difficult to predict which will significantly impact device performance in the field.

This desensitisation can be seen in the device literature, not solely for sunlight, but other forms of interference. Background modulated and unmodulated sources of radiation can have a similar impact on the detection capability and should be referenced in the test reports against standard test burns (5).

Exhausts from vehicles or power generation units like turbines can also interfere with the wavelengths at which Triple IR detectors operate. This can result in either desensitisation or false alarm, depending on the filter and sensor selected, and the nature of the algorithms on board the device.

The variation in performance to various test fires and environmental stimuli is unique to each device, which makes prediction of performance in the field challenging. Only if the designer has access to the proprietary algorithms and filter selection, we can predict the outcome of a test burn. Even then, however, we must still deal with the unpredictability of fire dynamics, particularly in an external setting (with varying wind direction, temperature, humidity, etc.) with such a high degree of potential flame/obstruction configurations, not to mention the variance in burning fuel and the unique response of each device to various fuels.

This device variation is demonstrated by Gottuk et al., who state 'all multi-spectrum IR OFDs do not provide equivalent performance' (1). These changes can be explicit and obvious such as the alteration of a Triple IR to specifically detect hydrogen (13), while others are more subtle, in which the causes of performance variation are not publicly available.

Visual Flame Detection (VFD)

Emerging in the late 1990s, visual flame detection is the most recent development in flame detection, with various forms of analysis and implementation available.

After experiencing the issues discussed in this chapter with the traditional detection technologies, the Oil and Gas industry, specifically Floating Production Storage and Offloading (FPSOs) facilities, required a detection technology which could be relied upon to reduce false alarms to flare reflections. The proximity of process relief flares to the deck on such facilities meant that traditional detection technologies experienced an unacceptable frequency of false alarm.

A technology was therefore required which could differentiate between a flame, and background-elevated readings of radiation emanating from the flare.

Visual flame detectors, as the name would suggest, operate within the visual portion of the electromagnetic spectrum. A technology based in the visual region allows the radiation which will encompass a facility to remain invisible to the detection, as it is to the human eye. While this is an over-simplification, it is often useful to conceptualise that visual flame detectors will only alarm to a flame which the human eye would recognise as a flame.

An additional benefit of looking in the visual region is that in addition to eliminating false alarms to 'invisible' radiation, the device will also not be desensitised to such radiation in the same way IR and UV devices would be (14, 15).

The camera-based nature of the technology allowed the advent of software-based analysis of the field of view. This enabled the development of algorithms which can recognise a flame and differentiate between a real fire and a false alarm source. This allows the technology to continually learn as the technology ages.

As with traditional detection techniques, however, no detection technology is immune from false alarm or desensitisation. The most obvious false alarm source is a real fire in the field of view (see introduction of this chapter). While this isn't really a false alarm as the device is detecting a real flame, poor design can result in a flare being present directly in the detector's field of view. This can also only present a temporary problem, which is not immediately clear at commissioning. Based on the inverse square law, the flare may only be large enough on certain days to result in an alarm. Despite the argument over whether it is a false alarm or not, the issue remains a critical consideration in device application and location selection.

A more reasonable argument of a false alarm source is that of a reflected fire on a perfect mirror-like surface. For example, the pooling of water which reflects a flare as a mirror image of the fire is a false alarm source. This would require removal of the source of the water pool, or relocation/reorientation of the device.

Remaining on the subject of water reflections, a traditional false alarm source for visual flame detection is the reflection of the sun when low in the sky, on water which is being disrupted (often by wind). This shimmering of the water, with the long-drawn-out sunlight reflection, can look like a flame to a camera-based technology, or the human eye. Fortunately, evolution of the human brain has allowed us to recognise the context of sunlight on water not to be a fire. Evolution of detector algorithms is still catching up.

This false alarm source, however, will not affect all detectors equally. As the technology is heavily software based, the technology is able to be 'taught' to distinguish between a shimmer of sunlight on water which may mimic a fire, and a real fire (while this can help speed along the evolution, it does not fool proof the detectors against future false alarms of this sort which mimic a flame source even more closely). The ability to achieve this false alarm immunity is a function of algorithm maturity and specific device manufacture.

Dependant on the version of visual detector selected, some devices provide a real-time CCTV video feed to the operator. As the visual technology is camera based, this provides the benefit of providing the video feed from the same lens as the detection device. This can assist first responders in dealing with a fire event, as they can view the area in real time as soon as the fire alarm is presented. This risk reduction to site personnel in controlling the situation remotely can be lifesaving.

As with all technologies, visual flame detection does have limitations on its use. The primary limitation is the inability to detect invisible flames. Any clear burning fuel such as methanol, hydrogen, and sulphur. simply cannot be detected by a visual-based technology (16). Generally, to once again use the analogy, if a flame burns and would not be detected by the human eye, the visual-based detector will be ineffective in detecting it.

With respect to desensitisation, the primary mode of reducing detection distance in the field is the impact of dirty optics. As with other detection technologies, if the window of a detector is dirty, this will reduce detection capability. This should, however, be reasonably accounted for during the design and F&G mapping, and through on-site maintenance.

In 2011, FM Global presented an independent review which resulted in the loss prevention data sheets (17). These recommended visual imaging flame detection for applications such as oil rigs. The rationale behind these recommendations were a result of the factors discussed in this chapter, but with the important point that each specific application requires analysis to determine what the best technology is for that area/hazard.

References

1. Gottuk D, Dinaburg J. Video image detection and optical flame detection for industrial applications. Fire Technology. 2012;49.
2. Talentum. FFE Broadspectrum Technology Method of Operation. Available from: https://ffeuk.com/wp-content/uploads/2021/08/Talentum-One-detector-for-every-application-Method-of-Operation-1.pdf.
3. Det-tronics. X2200 Ultraviolet Flame Detection Specification Data. Available from: https://www.det-tronics.com/products/x2200-ultraviolet-flame-detector.
4. Det-tronics. X9800 Single IR Flame Detector Specification Data. Available from: https://www.det-tronics.com/products/x9800-single-frequency-infrared-flame-detector.
5. McNay J. Desensitisation of optical based flame detection in harsh offshore environments. International Fire Professional. 2014;(9).

6. Chaplin M. Water Structure and Science 2000. Available from: http://www1. lsbu.ac.uk/water/water_vibrational_spectrum.html#d.

7. Det-tronics. X5200 UVIR Flame Detector Specification Data. Available from: https://www.det-tronics.com/Content/documents/X5200-Specifications.pdf.

8. Spectrex. Sharpeye 40/40l UVIR Flame Detector User Guide. Available from: https://www.spectrex.net/documents/guide-40-40-series-uv-ir-flame-detector-models-40-40l-lb-40-40l4-l4b-spectrex-en-us-1459808.pdf.

9. Det-tronics. X3301 Multi Spectrum IR Flame Detector Specification Data. Available from: https://www.det-tronics.com/products/x3301-multispectrum-infrared-flame-detector.

10. Micropack. FDS30 3Multi Spectrum IR Flame Detector Safety and Technical Manual. Available from: https://www.micropackfireandgas.com/flame-detection/multi-spectrum-ir-flame-detection-fds303.

11. Spectrex. 40/40i Triple IR (IR3) Flame Detector User Guide. Available from: https://www.spectrex.net/documents/guide-40-40i-triple-ir-ir3-flame-detector-spectrex-en-us-1459806.pdf.

12. Monitors G. FL4000H MSIR Flame Detector Performance Report. Available from: https://gb.msasafety.com/Fixed-Gas-%26-Flame-Detection/Flame-Detectors/FL4000H-Multi-spectrum-IR-Flame-Detector/pn/FL4000H-1-0-1-3-1-1-1.

13. Det-tronics. X3302 Enhanced Multi Spectrum Flame Detector Specification Data. Available from: https://www.det-tronics.com/content/documents/95-8779-2.1-(X3302-Pulse).pdf.

14. Micropack. FDS301 Visual Flame Detector Safety and Technical Manual. Available from: https://www.micropackfireandgas.com/flame-detection/intelligent-visual-flame-detection-with-video-fds301.

15. Draeger. Flame 3000 Visual Flame Detector Instructions for Use. Available from: https://www.draeger.com/Products/Content/flamme-3000-ifu-3360572-en.pdf.

16. Duncan G. Flame detector selection—Which one? International Fire Protection Magazine. 2018:72–4.

17. FMGlobal. Property Loss Prevention Data Sheets 5–48: Automatic Fire Detection. Factory Mutual Insurance Company; 2011.

5   Flame Detection in Process Areas

With respect to flame detection mapping, it is often striking how poorly understood optical flame detection remains within the hazardous industry. This is not at all the fault of practitioners, but rather the noticeable lack of research and peer-reviewed publications on the topic. From a personal perspective, I was very fortunate to work in an environment which provided me unrestricted access to a flame detection test ground and wealth of knowledge in the development of detection technologies. The vast majority of all I have learned on the topic has originated through practical testing on that facility and building relationships with those who have pioneered the flame detection industry for decades.

A lack of published data on flame detection performance and testing, however, can give way for the same mistakes of the past to be repeated. Consider a scenario where there is a seminar on F&G mapping discussing flame detection performance and mapping. The host presents on some fire tests that they have been able to participate in. A statement is then uttered in the form that seeing how flame detectors actually work was a real eye opener. This would be slightly concerning when also considering the scenario where software providers push the industry towards reliance on results of automatically populated detection layouts, when originators of such tools may not understand the fundamental performance capabilities of the devices. I'm sure that most would agree such a scenario would be alarming.

This may or may not seem farfetched, but I do recall a conversation with a former colleague who explained his shock when he heard a similar statement during a webinar, that understanding how technology works was not important in the placement of F&G devices.

This is not necessarily to lay blame for such narratives at the door of those stating them. We must consider if there is a failure in the industry in keeping test data, test processes, and detector performance characteristics in a closed loop within detector manufacturers and testing houses. A more open policy of research and journal publication, for example, may be best for the future of F&G detection design such that engineers and service/software providers offering related services do not go down dangerous paths in how the service or product is developed and promoted.

DOI: 10.1201/9781003246725-5

Designs of such systems require a deep understanding of how these devices work, how they differ from one another, and how the environment the devices are being placed in can affect the performance. Only then can we start to set our detection performance targets based on the risk and begin to analyse coverage effectiveness through application of software tools or other methods.

Practical Applications of Optical Flame Detection

As previously discussed, optical-based flame detectors are particularly effective in external environments where flaming fires are the predominant hazard. Let's examine why this is the case and why this technology is used in such applications.

For those who have seen an oil and gas installation, you will recognise the open nature of the structures, typically situated with the backdrop of a hostile environment. This environment can be harsh and unpredictable. Traditional smoke/heat detectors are therefore not an effective means of detecting flaming fires in such a location. Not only would the occurrence of faults be a regular issue (with smoke detection at least), but also the sensitivity would be extremely low. One can picture how developed a fire would have to become to activate a heat detector which is placed on the roof of a 10-m-high process module on an offshore oil rig in the North Sea. Even a smoke detector may struggle to detect the smoke from such a fire, assuming that the winds are strong in the area. This is a fairly reasonable assumption for the North Sea. Regardless of the weather conditions, such a fire would be well in excess of what would typically be an acceptable target fire size to protect the facility, the environment, and most importantly, its personnel.

When we consider the objective of the fire detection system in mitigating smaller fires from becoming larger fires, such as a major accident hazard (MAH), this demonstrates the application of traditional smoke and heat detectors as ineffective in such environments.

Considering that the objective is to detect controllable fires, this leads to the specification of how large a fire can safely remain undetected. This becomes the major input to the target fire size, which is the metric against which flame detection performance can be made.

This target has to be small enough to allow the mitigation measures to be effective in making sure the fire does not reach its potential. For this objective to be met, the designer has to make sure an appropriate target fire size is set for both alarm and control action (where detector voting is in place or differing levels of emergency response exist based on how many detectors are in alarm).

While having different target fire sizes specified for both alarm (where the fire can still be controlled manually) and control action (where the fire has developed to the point where operator personnel cannot be relied upon to control the hazard) has been present in operator guidance for decades, only recently has this appeared in an independent design guide (1). Legislation cannot be expected to go into the level of detail in specifying target fire sizes, but the absence of reliable independent guidance on the topic was of concern. This concern led to the development of a British Standard on hazard detection mapping, BS60080.

The governing regulation relating to F&G detection in the UK, 'Prevention of Fire and Explosion, and Emergency Response' Regulation 10 (2), states that the duty holder shall take appropriate measures:

(a) *with a view to detecting fire and other events which may require emergency response, including the provision of means for:*

   (i) *detecting and recording accumulations of flammable or toxic gases;*
   (ii) *identifying leakages of flammable liquids;*

(b) *with a view to enabling information regarding such incidents to be conveyed forthwith to places from which control action can be instigated.*

On the face of this, implementation of a simple approach of placing three flame detectors in a fire zone, operating in a 2oo3 voting configuration linked to executive actions, could comply with this requirement. When designing a detection layout for high-hazard industries, and a performance-based solution is sought, this can hardly be accepted as adequate. This task of demonstrating a system is fit for purpose is where flame detection mapping addresses the challenge.

Considering why flame detection mapping is preferable over the basic F&G philosophy statement of 'place four detectors per fire zone' begins with congestion. Where the area has congested regions, the target fire size may remain hidden behind such structures. Considering the scale of some fire zones, it is often unrealistic to assume that having only three devices would reliably result in two detectors detecting anything other than a significant fire in the area.

The initial effort at developing independent guidance on F&G mapping was the ISA TR84.00.07 (3). This guidance was an attempt at developing a framework for performance-based F&G Mapping. As such, the guidance was intentionally open, allowing designer freedom in implementation. The remit of the guidance was also to apply the performance-based principles of IEC61508/11 (4, 5), in the implementation of F&G mapping.

Initially, the focus pertained to that functional safety type safety engineering of calculating probability of failure on demand, system reliability, and theoretical calculations of detection success, for example.

Such efforts were admirable and broke new ground in the F&G mapping space, however, as with most performance-based design guides, it was still perceived in industry that designs could be in alignment with the TR while remaining inadequate in addressing certain specific hazards, as is the risk with performance-based guidance. The problem was therefore presented that we have Option 1: there is no independent guidance on F&G mapping, and design remains within a closed loop of experts; or Option 2: the performance-based design guide allows open access for designers unfamiliar with F&G, but familiar with functional safety, to begin designing F&G systems, with varying interpretations of the guidance.

This brought the issue of competence to the fore.[1] With the obvious drawbacks of maintaining an industry within a closed loop, the development of documents like TR84.00.07 is clearly the best way forward. This does require, however, those practicing F&G mapping to be able to demonstrate relevant competence in the field.

More recently, the aforementioned BS60080 aimed to provide a performance-based design guide for hazard detection mapping independent of the functional safety philosophy. This guidance aimed to reduce the potential scope for error by focusing primarily on the practicalities associated with F&G design, providing guidance on design methods as well as the technologies themselves.

The issues surrounding relevant experience are not trivial. Where a designer does not have experience with the technologies being applied, the philosophy behind the mitigation action, or the methodologies being applied, for example, there are areas where critical shortcomings can be implemented in design. The extremes of 1) using a software tool blind to design the detection layout automatically, and 2) using only engineering judgement with no numerical assistance are each unlikely to generate an optimised design.

Take detector effectiveness, for example. Can a designer adequately map a suitable detector layout if they are unaware of the specific capabilities and limitations of the device? Can the design adequately address the manufacturer claims of performance? Can we verify these claims are applicable to field-based operation of the device?

These queries in no way suggest that detector manufacturers are misleading in their detector datasheets. This is purely to highlight that we cannot possibly gain all the information we need from a data sheet or user manual. Flame detectors will be capable of detecting the fuels specified on the data sheet or manual, otherwise they would not achieve certification. The designer, however, is required to interpret further than what is presented. The designer must consider the environment the detector is being placed, the congestion in the areas, the hazards present, the escalation potential, and also the variability in fires dependent on a multitude of factors (the same fuel being burned in the same manner can appear totally different on two separate days).

When considering response time, for example, the design must consider acceptable thresholds.

Response times of <10 seconds are commonplace in the hazardous industries; however, one must note that devices may have varying response times across the claimed FOV. For example, a device may have a delayed response when the flame is burned at maximum off-axis, compared to a fire in the centre line of the detector.

Considering again the environment that the devices are located, it must also be noted that the devices a manufacturer sells to the Sahara are likely the same devices they will supply to the Alaskan Prudhoe Bay. You also have facilities like the BTC pipeline which experiences each of those extreme conditions as the seasons pass. These devices are not replaced for winter/summer as the environment changes; therefore, performance variability must be considered.

An extreme example would be an artic site which applies a detection technology which works well in the summer environment but is sent into alarm when the snow falls in winter, causing a flickering of the hot processing equipment the detectors are looking at. Facility operators can't simply turn the detectors off for winter.

With flame detectors of all technologies aiming to detect sources of radiation which exist in our everyday world, the potential for false alarm can't be eliminated, as discussed in Chapter 4. When one considers that the natural light across our planet is generated by a giant fire, the issue of false alarm is one that flame detectors will be wrestling with for some time.

These discussion points on the environment, response time, congestion, etc. are the tip of the iceberg of what must be accounted for during design. For this reason, it is clear that we require more than a software package to design such systems, and competence in the relevant field is critical. For this reason, Chapter 9 will discuss competence, specifically within the context of F&G mapping.

Reverting to the introductory anecdote claim that 'it doesn't matter how a flame detector works when it comes to mapping', I would hope this brief introduction to the practical application of flame detection shows the incongruity of such a suggestion.

Performance Testing

Obtaining an independent overview of detector performance is critical to assist with both detector selection, but also mapping. Such an independent review can be provided by certification bodies such as FM Global, who test devices against recognised performance standards (7, 8). Designers and procurement can therefore review the test results as tested by FM, or equivalent, to review a comparison of performance between detectors using a level playing field. This is assuming that such documentation is provided by the manufacturer. The absence of such information from the manufacturer can itself be a useful performance indicator.

While manufacturers are under no obligation to undergo such testing or provide the results, the results which are presented in the test reports are important for a designer to understand the device performance. It is also important for the designer, however, to understand that the detector is only tested against certain controlled fires. This cannot possibly represent the full and complete performance of the device for all applications.

In analysing performance of a flame detector, there are a number of different performance tests which are of interest.

The first test is that of baseline sensitivity, which determines the range of a flame detector to a specific fire in optimal conditions. The standard test is traditionally a 1 ft² n-heptane pan fire, however, baseline sensitivity tests against other flames are also common (Methane, Propane, Diesel, and Methane, for example). While these tests are a limited sample of the detector performance, and a fire in reality is unlikely to mimic these fires, they give a good comparison between devices.

Another set of test results which can be of interest is that of the response to false alarm sources. Such sources of spurious alarm can include radiative sources (both modulated and unmodulated), sunlight (both direct and indirect), welding, etc. These environmental factors can be reproduced to verify performance with respect to false alarm rejection, but they can also be presented to check their effect on flame detection performance (i.e. the level of desensitisation they provide).

The specific cone of vision or field of view is another factor which is revealed in such tests. This provides the detector's horizontal and vertical field of view, along with the extent of the reduction across the field of view.

Figure 5.1 (9) shows these factors affecting the selection of a flame detector.

One significant performance influencing factor which is less likely to appear in the test report is the performance of a flame detector in the presence of congestion in the field of view. Such congestion can be a horizontal

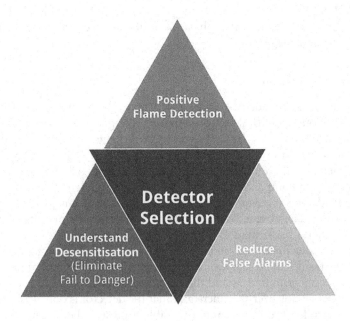

Figure 5.1 Flame detection selection.

or vertical blockage, or combination of both, in an almost unlimited variance of configurations. This makes detection 'prediction' almost impossible, something which will be discussed later in this chapter. This issue is further exacerbated by the previously discussed unpredictable environments that these devices are placed within (10), often with a significant effect.

The available technologies have been discussed in Chapter 4, but each technology has a wide range of devices which operate using that detection method. Where two different devices from the same technology are tested side by side, their performance to various fires can vary significantly. The performance of each device and the subsequent delta between them can differ based on changes in environment, blockage, burning fuel, and so on. It is not credible for a test process to account for all such factors, meaning that these performance variations remain unknown for the most part.

Even when devices have a clear line of sight to the flame, performance can vary between test burns of the same device, with no variable introduced which one would expect to affect detectability (11). Designers can review coverage and have a knowledge of the devices themselves, but whether detection will occur or not will remain unknown. It is therefore important that flame detection mapping presents the true coverage (clear fields of view) afforded by the detector, such that adequate engineering decisions can be made, rather than the presentation of a predicted detection which, as demonstrated, can only be flawed in its assumptions. This flaw emerges either by applying assumptions of detection capability, or not having access to the specific filter selection and on-board algorithms of detection decision making of the devices, coupled against the knowledge of the flame and its behaviour in those wavebands.

It is responsible practice to include an appreciation of this test data in any mapping study to ensure that the accurate detection performance is mapped for the specific detector selected.

Table 5.1 demonstrates the performance implications of the various environmental sources listed, which would be expected from a standard Triple IR detector. Note that this is not representative of a specific device, and the specific test data must be sought when reviewing detector performance.

This benchmark performance testing is beneficial in the flame detection industry. The benefits are extended to detector suppliers who can show off their detector's performance, those procuring the detectors, and those designing F&G systems. The data allow engineers to implement more accurate readings of detector capability. While the tests do include environmental testing similar to those in which the detectors are being placed, it should always be noted that they are not a representation of what that detector will achieve on the day of a fire.

To once again draw a parallel with Fire Safety Engineering, this issue is similar to the problems noted around the controversial but enduring facet of 'fire resistance' of construction materials (12). With a fire resistance rating of, for example, 60 minutes under a standard fire test, many assume that

Table 5.1 Detector Performance Implications to Environmental Influencers

Environmental Stimuli	Test Fire	Detection Distance	False Alarm (Y/N)
Indoor optimal conditions	1 ft² n-heptane pan fire	40 m	N
Direct sunlight	1 ft² n-heptane pan fire	20 m	N
Direct modulating sunlight	1 ft² n-heptane pan fire	20 m	N
Indirect reflected modulating sunlight	1 ft² n-heptane pan fire	30 m	N
2 kW electric bar heater, steady state	1 ft² n-heptane pan fire	10 m	N
2 kW electric bar heater, modulated	1 ft² n-heptane pan fire	N/A	Y

this equates to a guarantee of survival to any fire for at least that duration. In particular, if that material is a structural material, it should be capable of withstanding and surviving burnout. This presents a lack of understanding of fire dynamics [particularly when considering timber frame structures (13)] and is an assumption made on the basis of a limited, and often criticised, test. Such criticisms include the lack of appreciation for the decay phase of the enclosure fire. The fire resistance test, however, is not there to show what will happen during a specific fire, but rather to present a standardised test of the material, the results of which fire engineers can interpret in designing a building.

This parallels well with the flame detection performance tests. These tests do not guarantee that a device will perform on any given day to how it tested on the day of the test. The results of these tests are a record of how the detector performed on the day of that test, to that specific fire. It is the designer's responsibility to use this information to design a suitable detection layout.

Implementing Flame Detection Desensitisation in Mapping

In order to implement flame detector desensitisation into the flame detector mapping process, the initial baseline viewing distance to a standard test fire is obtained (as a result of the previously discussed testing procedure). While this standard 1 ft² n-heptane pan fire is not representative of a potential fire which may occur in every area, it is useful in showing a detector's capability to a typical hydrocarbon pool fire. Note that care should be taken to ensure that the value of this baseline sensitivity aligns with the sensitivity setting intended to be used. The maximum capability of a detector set on a high sensitivity may be promoted by a supplier but may not bear a resemblance to the setting supplied in reality.

This baseline sensitivity of the selected device becomes the starting reference point in analysing detector capability. The designer can then apply the data from the literature of the detector's capability in the presence of environmental stimuli known to cause false alarms and desensitisation, including sunlight, dirty optics, and modulated and unmodulated radiation in the FOV (9, 14). This altered baseline sensitivity is often referred to as the 'effective viewing distance' and provides a more reasonable performance of what the detector may achieve in the field.

Three primary factors to be considered in the calculation of flame detector performance in unpredictable environments, which are well documented in the literature (9, 10), are presented later.

1) Performance impact when exposed to spurious stimuli

The detector literature, including technical/user manual and datasheet, should normally provide the designer/procurement with data on the detector's performance in the presence of the most common false alarm sources (background radiation, direct/reflected sunlight, etc.). The testing should state the proximity of the false alarm source to the device, and the detection distance to the particular fire which was achieved. As previously mentioned, if data from such fire tests are not readily available, this may indicate poor performance, but this is not always the case. Discussions with the detector manufacturer are always important. Where such specific data are not available, a suitable value for desensitisation as a result of these common false alarm sources should be applied.

2) Dirty Optics

Something that all optical-based flame detectors are affected by is the issue of dirt in the optical window. As environmental dirt and grime gather on the detector window, the sensitivity to a flame will reduce. Obviously, we do not want a device to register a fault when the slightest bit of dirt is on the window (as maintenance technicians would essentially become full-time window cleaners), but we do want to get a fault from a device when it is inhibited from detecting a fire.

For UV and IR detectors, a typical threshold of reduced sensitivity which would register a fault is 50% reduction. In theory, this means that in the field, a detector could have a reduced sensitivity of 51% without registering a fault. Such a reduction would be overly conservative to assume during design, just as assuming that 100% clear windows would be overly optimistic. A reduction of sorts will need to be considered. It should also be considered that with dirty optics, the grit or dirt is not generally uniformly distributed across the window, and this non-uniform distribution of dirt can affect detector performance, and performance effectiveness of the optical verification/integrity test on some devices.

This desensitising factor also highlights the importance of maintenance checks of the devices. Operators should not wait until an optical fault is received, and should ensure that visual inspections are carried out as required by the manufacturer. Guidance on maintenance intervals is also provided by FM Global (15), who recommends that IR and UV detectors are tested every six months, and visual-based detectors are tested annually. Individual events can also trigger a test, however, such as storms which have the habit of coating detectors in optical contaminants.

Regarding the testing, a suitable method of testing is to use a test torch from distance. This has the benefit of checking whether the optics within the device are working, but also to make sure that the window is clear enough that it can see the area being viewed.

3) Impact of filter edge effect

The third facet of flame detector performance is that of sensitivity across the claimed field of view. Typically, the sensitivity reduces as the flame is positioned away from the centreline of the field of view. This reduction in sensitivity is normally the result of the filter edge effect.

The severity of this decrease in sensitivity is unique to the specific device (typically because the specific filters will differ).

Figure 5.2 shows a comparison of flame detector cones of vision between a visual flame detector (which suffers less from sensitivity reduction off-axis) and a Triple IR detector.

While the sensitivity of visual flame detection is generally reduced at extreme off-axis, this is not for the same reason as a radiant-based device. With IR and UV detectors, it is not uncommon to see a reduction of 50% of the centreline detection capability off axis. With visual detectors, the reduction in sensitivity relates primarily to the portion of the flame which burns outside of the detector's field of view. For example, when a fire is at the extreme off-axis, a portion of the flame will begin to dip outside the camera's field of view.

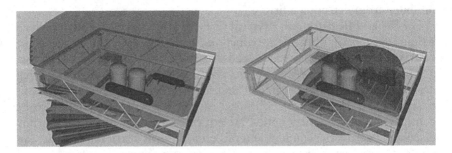

Figure 5.2 Visual flame detector cone (left) vs triple IR flame detector cone (right).

Figures 5.3 (16) and 5.4 (17) present example detector footprint shapes. These show, particularly in Figure 5.3, the impact of the filter edge effect. Note that many more manufacturers and models of each of these technologies are available at the time of writing.

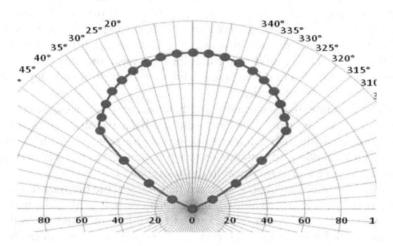

Figure 5.3 Honeywell FS24X.

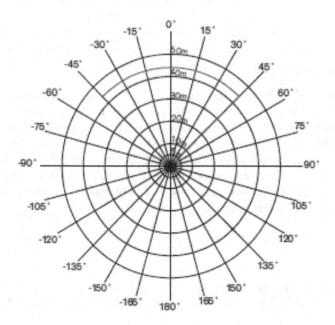

Figure 5.4 Dräger flame 5000 (14).

It is important during the mapping phase to ensure that the filter edge effect has been adequately addressed in the representation of the specific flame detector footprint.

When we consider a typical application in which these devices are installed, it is simply not credible at the time of writing to include all possible environmental factors during design, and also to ensure that such factors are relevant at all times. Consider, for example, an FPSO processing deck, with exposure to the environment offshore, as well as the variable radiation from the flare stack. The variability in conditions will mean that any flame detection devices installed in this area will have varying performance day to day, through the life of the facility.

This variability and uncertainty are therefore generally accounted for by determining the baseline sensitivity, and applying relevant values for the three previously discussed desensitising factors for that specific application in the understanding that this is only a benchmark level of performance. Engineering judgement is critical in this process.

Where it can be reasonably assumed that the specific fire hazard will be known, the performance value to that fire may be best used as the baseline sensitivity value, rather than the standard 1 ft² n-heptane pan fire. As flame detectors can have varying performance characteristics based on the burning characteristics, this can be an important value in the accuracy of the design.

When using this value, however, the designer must be aware that the test data on desensitisation is generally to the 1 ft² n-heptane pan fire. These values may be inappropriate based on a different burning fuel. Liaising with the manufacturer and engineering judgement is once again crucial here.

Calculation of RHO

While this book will generally discuss the 1 ft² n-heptane pan fire as ~40 kW RHO, there are various calculations and debate in the industry as to what RHO this fire produces. Designers must therefore be cognisant of what their baseline fire size is before applying risk-based target fire sizes. The following shows one example calculation of RHO of the 1 ft² n-heptane pan fire.

To calculate the heat release rate of a 1 ft² (0.0929 m²) n-heptane pool fire, the following equation, for pool fires >0.2 m in diameter, is used from SFPE *Handbook of Fire Protection Engineering* (18):

$$Q = m'' \Delta H_{c,eff} (1-e^{-k\beta\,D})\, A_{POOL}$$

where:

Q = pool fire heat release rate (kW)

m'' = mass burning rate of fuel per unit surface area (0.101 kg/m² sec for n-heptane)

$\Delta H_{c,eff}$ = effective heat of combustion of fuel (44,600 kJ/kg for n-heptane)

A_{POOL} = surface area of pool fire (area involved in vapourisation) (0.09290304 m^2)

$k\beta$ = empirical constant (1.1 m^{-1} for n-heptane)

D = effective diameter of pool fire (diameter involved in vapourisation) (0.344 m).

Noting that the equation assumes complete combustion, the total HRR is ~130 kW. To calculate the subsequent portion of radiative loss, this is estimated at approximately 30% of the total HRR for pool fires of this size (19).

Therefore, using the HRR value and assigning the portion of this as radiative loss as discussed earlier, we can solve the following:

$$RHO = Q*XRA$$

where:

RHO = radiant heat output (kW)

Q = pool fire heat release rate (kW)

XRA = radiant fraction of total heat release rate.

Using ~30%, this is roughly equivalent to 40 kW RHO for the standard baseline pool fire applied in flame detector performance testing.

It is important to note this does not account for the multitude of errors which can influence the burning rate, including wind speed, direction, and temperature and is entirely an estimate of RHO under ideal test conditions.

The value of RHO can be corroborated by calculating the theoretical flame height and verifying if this is reflected in ideal test conditions. Using the SFPE Handbook (18) once again, the following equation shows the calculation of flame height, as developed by Heskestad:

$$LF = 0.23Q2/5-1.02D$$

where:

LF = the 50 percentile intermittent flame height (m)

Q = the heat-release rate (kW); taken from answer above = 131.82 kW

D = the diameter of the fire (D); taken from above = 0.344 m

In solving the equation, we see a transient flame height of 1.27 m. As this is reflective of the flame heights witnessed in practical testing, this adds credibility to the results of the RHO calculation.

Flame Detection Performance Targets

Arguably the most critical point in performance-based flame detection mapping is the setting of the performance targets. This process gives the designer the ability to optimise the layout of detectors. Set the target too small and the design will be overly conservative. Set it too large and the design will be unsafe. That goldilocks zone is where optimised detection occurs. Regardless of software applied or algorithms used in detector placement, the setting of the target is one of the most important areas of mapping.

It is also one of the most controversial areas in detection mapping. Before considering how to set the targets, another representative scenario will be presented to demonstrate how the setting of performance targets may not be completely fool proof.

Picture the scene. An onshore facility has applied F&G mapping software to verify the placement of flame detectors. The facility stipulates a simple uniform target percentage coverage which has to be achieved for the hazardous machinery/process equipment. This is generally not an optimal method of designing an adequate system where there is variance in the risks present, and this scenario aims to substantiate this critique.

The facility is modelled using volumes around each piece of equipment within which a target fire size of 250 kW must be detected. The acceptance criteria require 80% of said volume to be directly visible to the flame detectors to achieve 'adequacy'. The graded equipment is surrounded by open space and the detectors are positioned away from the equipment. The resulting assessment does not achieve the target percentage of coverage. Rather than altering the detection layout, or justifying the deviation from the arbitrary target percentage coverage through engineering judgement, the designer simply extends the graded volume further out from the equipment into open space. The result of this is a greater percentage of the graded volume being visible to the flame detection, as there is no obstruction, with no risk present in that new extended location.

The resulting assessment shows that the acceptance criteria are achieved. Consider this again. The initial assessment revealed that the detection was inadequate against the pre-set acceptance criteria, because it didn't achieve the target percentage. Nothing physically has changed to the layout and the risk has not changed, but on the next assessment run, the coverage is adequate, and the design is approved.

This is not to suggest that a design alteration would be required, but rather: why do we have the target percentage in the first place? If performance-based design and the setting of performance targets are to be successful in flame detection, then guidance, standards, and designers must avoid the setting of arbitrary figures of 'adequacy'. They must apply engineering judgement and competence to the output of a numerical analysis to determine coverage adequacy.

Moving on to the process of setting target fire sizes and acceptance criteria, these performance targets, or grades, must provide a stated target fire size (typically in kW RHO) which is the maximum allowable fire size which can remain undetected. When a fire reaches this size, it must be detected. A target threshold for voting can also be specified where voting for automatic executive actions is present. This second target fire size is typically a larger fire, as this will result in automated action. This should typically not happen unless the situation is beyond the point of manual intervention, or unless the risk is high enough to warrant executive action of a small fire.

As a point of note, RHO is not always the metric applied in the setting of performance targets. Effective viewing distance (D) can be applied to simplify the targets. This essentially models a detector's detection capability to a predetermined fire, and uses multiples of that value.

Tables 5.2 and 5.3 show example values which could be applied for various facilities. Note that these are not specific to a site or specific guidance document, but are purely representative of the type of values common in flame detection mapping.

As one can see from the values stated in Tables 5.2 and 5.3, the target fire sizes are not those associated with the expected or anticipated fires from high-hazard pieces of equipment. Such fires would be expected to potentially reach the MW level of RHO.

The values presented in the tables are a result of a flame detector's detection capability being based on the inverse square law. Essentially, a flame detector has an infinite detection capability, assuming that the fire is large enough. Similarly if the fire is small, the flame detector may still detect the

Table 5.2 Example Offshore Hydrocarbon Risk Area Grades and Associated Fire Sizes

Risk	Fire Size (RHO) Alarm	Fire Size (RHO) Control Action
Extreme	10 kW	10 kW
High	10 kW	40 kW
Moderate	40 kW	160 kW
Negligible	160 kW	320 kW
Unique	To be defined if none of the above is suitable.	To be defined if none of the above is suitable.

Table 5.3 Example Onshore Hydrocarbon Risk Area Grades and Associated Fire Sizes

Grade	Fire Size (RHO) Alarm	Fire Size (RHO) Control Action (where applicable)
Extreme	40 kW	40 kW
High	40 kW	160 kW
Moderate	160 kW	320 kW
Negligible	320 kW	640 kW
Unique	To be defined if none of the above is suitable.	To be defined if none of the above is suitable.

fire, but it would need to be much closer. With this in mind, where a detector can detect a 40 kW fire at 30 m, the same detector should be capable of detecting a 10 kW fire at 15 m. As the RHO quarters, the detection distance halves. Similarly, a 160 kW fire would be detected at 60 m. When the detection distance is doubled, the detector required four times the RHO to provide a comparable response.

The target fire sizes instead target small fires from external sources or adjacent pieces of equipment which could cause escalation from that targeted piece of equipment. It is often a surprise that the targets set for a piece of process equipment are generally not to detect fires from those vessels, but rather to detect fires from elsewhere which are in the vicinity of the targeted equipment. Modelling or targeting anticipated fires from such pieces of equipment, which becomes largely irrelevant in this respect.

In addition to the types of risk grades specified in Tables 5.2 and 5.3, specialised risk can also be present. For example, a methanol tank may present a Moderate risk in its given location, but only certain variants of UV and IR detection can detect such a flame. It would be pertinent therefore to include such a statement in the performance target definition. Equipment like this can therefore be assigned as a special risk, with the performance target and acceptance criteria specifically defined.

Flame Detection Mapping

Once the performance targets are set and the acceptance criteria agreed, the coverage of a preliminary or existing layout requires to be mapped to verify the target fire sizes will be detected (1). This process pulls together the considerations discussed in this chapter.

The application of mapping tools/software emerged in the 1990s and early 2000s (20, 21) and is a fairly standardised process in the hazardous industries today (22–24).

The primary role of such a tool is to model a representative device capability against the predetermined performance targets and provide the engineer with a coverage representation of the area. This coverage representation is then reviewed by the engineer to decide whether the coverage is inadequate, optimised, or over engineered.

As this chapter has demonstrated, however, this modelling phase is a small part of the overall process. Today, a significant emphasis is placed on the mapping software selected, as this provides the graphical output which generally attracts attention.

Such a graphical output is critical in assisting the detection engineer, but arguably of far greater importance is the multitude of facets previously discussed. It is always important to remember that the software is a tool to be used to assist engineer decision making, not to do the job of the engineer.

The process of flame detection mapping typically begins with the 3D representation of the facility. Once all of the processing equipment is present, the risk-based grades are applied. The proposed or existing device locations are then selected, and the detectors are added into the scene.

The tool can then run analysis of whether the detectors are able to detect fires of the specified size, within the volumes specified. If the resulting detection is not adequate, alterations can be made. Equally, if the detection seems excessive, the optimisation process can begin.

Figure 5.5 shows a typical flame detector location, with the specific device cone of vision projected. The cone of vision can be seen to be obstructed by equipment in the scene. Congestion within the volume is one of the primary reasons that flame mapping tools are particularly useful.

Figure 5.6 (22) shows a typical 3D environment with various risk grades applied in the area. Each coloured 'bubble' corresponds to a target fire size. Where detector voting is included, these grades often have two target fire sizes included—one for alarm and a larger target fire of automated control action upon alarm from a second detector.

In its most basic form, flame detection mapping can provide coverage analysis based on a single target fire size or detection cone (i.e. a 40-m detection cone) per grade or volume. This tells whether the location is visible to no detectors, one detector, two detectors, or more than two detectors.

This analysis may only be suited, however, to low-risk applications and may not provide a truly optimised performance-based design.

More enhanced flame detection mapping software can accommodate multiple risk grades in a volume, each of which can have an individual target for alarm, and a different target for executive action (such as the targets for

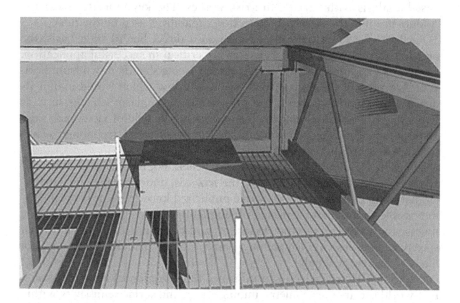

Figure 5.5 Obstructed 3D flame detector cone.

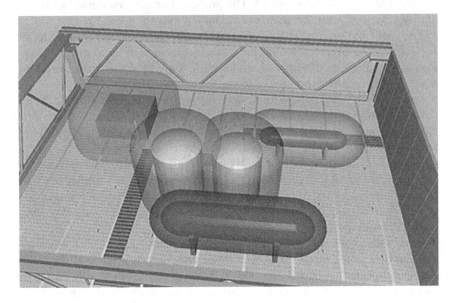

Figure 5.6 Typical risk grading representation (17).

High, Moderate, and Negligible grades listed in Table 5.2). Such an analysis automatically applies the inverse square law to the coverage analysis.

Figure 5.7 (22) shows a 2D slice of a 3D analysis which includes a performance-based analysis with varying fire risk grades. The key indicates what the colours mean in the context of detection capability based on the risk.

As flame detectors typically operate on a direct line of sight basis, the positioning and orientation of devices are critical in successful application. Devices, for example, should be positioned facing slightly down from horizontal to prevent build-up of dirt on the window which would inhibit its detection capability. Equally, the presence of false alarm sources such as flares should be ensured not to creep into a device field of view. Such false alarm sources, and more importantly incorrect detector technology selection, are a significant cause of false alarm.

When the results of the detection are generated, the engineer must still apply their knowledge of the area, the hazards, and the detection capability. Based on current and foreseeable publicised knowledge, it is practically impossible to automate a means of simulating the likelihood of detection unless you have access to the algorithms and signal processing used by that specific flame detector (something flame detector manufacturers are not likely to present willingly).

Additional factors which can influence the detection characteristics in a real-world fire are the geometry, impingement, and surface emissivity within the area the fire occurs. Such factors can be challenging to account for within

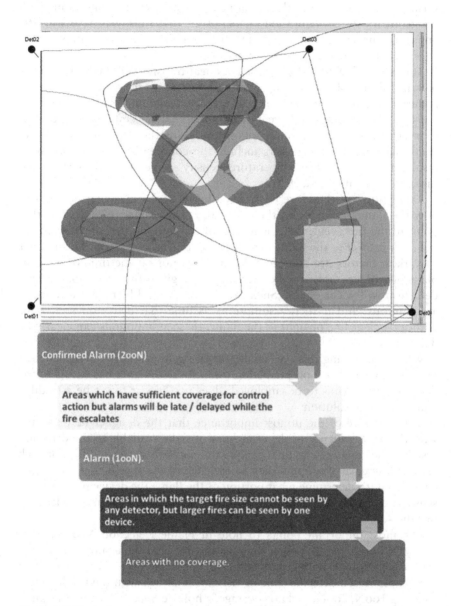

Figure 5.7 Typical assessment output of flame detection mapping (17).

design. Even more challenging, these can be incorporating into an analysis of 'successful detection' versus 'failure to detect'.

Consider, for example, the surface emissivity of materials within the burning area, i.e. the effectiveness with which a material emits thermal radiation. As the fire begins to impact equipment within the fire area, the surface emissivity of both the radiating and receiving materials, along with the geometry (or configuration factor) of the materials, will have a drastic impact on the nature and extent of the RHO. This will therefore impact how effectively the flame detectors will perform.

Annex C of PD7974-3 (25) discusses heat transfer and thermal response of materials and shows the impact of such factors on the temperature rise of materials in a fire including concrete, protected, and unprotected steel. Such a rise in temperature will have a direct impact on detectability with respect to interference on the detection algorithms along the electromagnetic spectrum. Such influencing factors include the conductivity and thickness of insulating materials, fire temperature, density of the material, specific heat capacity of the material, etc.

This value will therefore be affected by what the receiving material is, how the flame impinges on the material, the dynamics of the fire at that point in the fire event, the orientation of the burning fuel onto the material with respect to the flame detectors themselves, and so on. The number of variant factors are considerable, and it is not (at the time of writing) plausible to *accurately* predict, for a specific given fire scenario, whether detection will occur or not. Such calculations would be required to make assumptions, and if such assumptions assume that detection will occur without a clear line of sight to the flame, these assumptions can clearly be dangerous.

When considering this one small area of fire dynamics, it becomes clear that automated detection prediction (even if the designer has access to the detection algorithms to be employed) does not appear to yet be a credible and safe design solution.

It is therefore of the utmost importance that the designer be presented with the hard evidence of the clear lines of sight available to the detection system. The engineer can then analyse the performance target fire for each area of low coverage, apply their knowledge of the specific detector being considered (to determine what portion of the flame the device may be more sensitive to), and analyse the nature of the blockages to determine adequacy or otherwise.

The most important points to note in Flame Detection Mapping are: always account for blockages and shadowing; and understand the design aspects other than simply reviewing percentage coverage of the area. Often if a simple closed detector cone of vision is analysed with the result simply showing 1ooN, 2ooN, and no coverage, a holistic approach cannot be analysed, and will not provide the designer with sufficient information in determining flame detection adequacy for high-hazard applications.

As with a variety of engineering disciplines, practical test data are critical in understanding how best to employ the detection system.

Figures 5.8–5.11 (9) show various configurations of test set up used to understand the impact of congestion on a flame detector's capability to detect fire.

Figure 5.8 Example congestion configurations to various test burns (12).

Figure 5.9 Standard fire size viewed though vertical and horizontal pipe obstructions (12).

While these examples are limited and barely scratch the surface of the number of possible configurations, fire types, and so on, they are useful in demonstrating the variance in detection results even between two devices of the same technology (9).

Results vary considerably from device to device, even within the same technology band. As an example, two Triple IR detectors can provide wildly different results when exposed simultaneously to the same fire and blockage.

Figure 5.10 Piperack obstructions with radiating blackbody in the field of view (12).

Figure 5.11 Angled piperack in the vertical position simulating up to 50% obstruction (12).

The science behind fire dynamics is highly complex in nature. It is for this reason the application of detection design principles and analysis of adequacy be lent the same degree of weight applied in other life safety disciplines. Considering specifically the mapping software applied, it is critical that such a tool is designed to comply with the philosophy in which detection is designed against. With limited validation available, it is critical for designers to adequately judge if the tool being applied is fit for purpose.

Putting the Design Engineer at the Heart of Flame Detector Optimisation

The considerations discussed earlier will influence the design engineer's analysis of adequacy. Such modelling is intended to show the designer the coverage of the volume, either by basic analysis of 'coverage' or 'no coverage' within a cone (Figure 5.2), or by adding complexity in modelling the specific detection capability to various target fire sizes as shown in, for example, Figure 5.12 (applying risk-based techniques to coverage analysis) (9).

One significant factor in detection adequacy is to ensure that the design is not reliant on the cusp of the margins. For example, as there are so many fundamental issues associated with flame detection reliability, a suitable

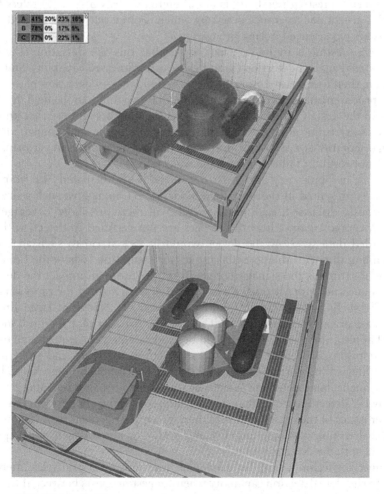

Figure 5.12 Risk based flame detection coverage assessment.

design would not rely on detection of a single device at the marginal point of the detection footprint, i.e. maximum off-axis at maximum detection distance. Such factors include, but not limited to, environmental influencers (rain, solar/blackbody radiation, equipment blockages, etc.); obscuration of the optics; installation/commissioning accuracy issues (i.e. you cannot be sure the device is orientated looking directly where it was intended); manufacturing tolerances of orientation; design placement of the device orientation from layout generation (for example, it is easy for a device drawn onto an F&G layout to be off by a few degree—this alteration across an 80-m detection distance can result in a large deviation). Tolerance issues like this (either from manufacturing [although if SIL, for example, is applied, one would expect much tighter tolerances], design, installation, and environmental) are currently unavoidable. A typical threshold of +/-5% acceptable tolerance on testing reliability is often referenced. EN 54–10 (8) refers to a +/-5% acceptable tolerance in testing which is often referenced when deviations are experienced during fire testing.

As fire testing is by its very nature variable, this 5% deviation could be considered against detection claims/previously witnessed capability. Suitable design, therefore, must ensure sufficient redundancy in detection placement and procurement to account for this potential variability in the field. It is the role of the design engineer to be aware of such factors and use that knowledge to determine adequacy such that a variation in the environment, detection orientation, or any of the previously discussed variances in performance does not result in failure of the flame detection system.

Flame mapping may also look to simulate the 'anticipated' fire from the hazards presented in the area, then model detection against such a fire. As previously discussed, such an approach will therefore show 'coverage' or 'no coverage' against fires the devices are not certified to detect, with no recorded data to show the device would in fact alarm to such an event. Coupling this with the potential for sensor saturation (where the fire is so large it saturates the signals received from the fire) will impact the devices in unique and unpredictable ways, for example, distorting the ratio reliance in a Triple IR and potentially 'blind' the device (11). Note that 'large' fires in this case are approx. 600 kW RHO, somewhat smaller than a typical industrial application flame which would be modelled in a consequence analysis. Assumptions, therefore, surrounding detection capability and subsequent modelling of such fires (for example, using CFD combustion models to base subsequent flame detection designs against) may present misleading results. Only UV detectors are immune to such unpredictability, but present multiple other issues of their own. It would therefore appear evident that designing a flame detection system to target such large fires is not adequate when applying most flame detection technologies.

Emerging techniques in flame mapping draw on similar principles, attempting to simulate detection capability when congestion exists between the flame and the device (26, 27).

Such an approach must consider assumptions made on plume behaviour, environmental conditions, detection algorithms, filter selection of the device, detection technology, and other factors as previously discussed. Validation of small bore pipework between the device and a fire close to the device will provide positive results where the software has registered a positive detection capability, but: 1) this would not be acceptable on a facility with a high congestion factor or larger areas, and 2) designer guidance does not require the addition of additional devices as a result of such minor blockages in a small area when using an approach which represents the true clear coverage from a device (1, 28). Compounding the issue further is where the operation of the flame detector is software-based (as is the case with most detection), and therefore subject to change as the algorithms mature. This may therefore require update to the software tools each time the device firmware updates (11). This may also mean any facilities which have applied the approach may have to revalidate their flame detection mapping when the device firmware is updated on site.

The orientation of the device is also critical in such an approach, the impact of which will vary from device to device (11). For example, optimal results in a test burn may be achieved by placing a detector at a low level looking perfectly horizontally at an elevated fire, but this is not reflective of true application and detection performance can therefore be misleading. Understanding these nuances is critical in detection placement. Detection performance can also be artificially improved during testing if the device is set to its highest sensitivity (which may not represent what would then be applied in the field due to excessive false alarms). Performance can also be improved where the devices are shielded from environmental factors like sunlight.

There is a fundamental issue, however, as previously discussed that it is currently not possible to incorporate the proprietary detection capability of all flame detectors in modelling software. Assumptions will be inaccurate if applied to all devices which operate using different technologies, filtering, algorithm analysis techniques, and where environments are changeable, as shown in Figure 5.13.

The literature (10, 29, 30) tells us flame detectors can be desensitised by sunlight, optical contamination, and modulated black bodies to name but a few external influencers. We therefore need a way to use the data from FM3260, along with other factors, to provide a fair comparison between coverage of detectors as discussed previously.

The other critical factor in determining coverage adequacy is the presentation of a target percentage coverage. This is, however, not a good metric of adequacy (31). BS60080 (1) demonstrates that this with the 'chessboard' analogy, as shown in Figure 5.14 (1), whereby 50% coverage of an area may appear like a chessboard or a 50/50 split through the middle of a volume. It is unlikely a design engineer would review both as 'adequate'.

There exists significant uncertainty in this process. As discussed in BS7974 (32), the difference in radiant heat flux calculation based on the specific nature of the fire, with the subsequent reliance on an accurate measure of

Figure 5.13 Environmental variance which will impact detection behaviour.

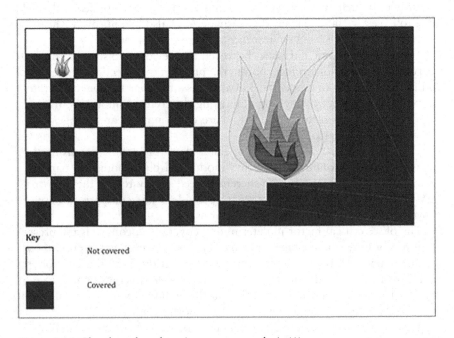

Figure 5.14 Chessboard analogy in coverage analysis (1).

flame temperature can be drastic. In certain calculations of radiant heat flux, a simple 10% alteration in flame temperature (i.e. 1100 K rather than 1000 K) used in the calculation can produce a 46% error in the resultant radiant heat flux calculation. Where this radiant flux is used in determining flame detection success or not, this is not a trivial alteration.

When one considers further that such fires are expected in external areas with all of the unpredictability discussed previously (in both this chapter and Chapter 4), the idea of using algorithms to predict flame detection success or otherwise is discredited further.

It is pertinent, therefore, that designers are required to analyse the performance of the flame detection layout (output of the software analysis if required), the nature of blockage, the technology applied, the hazards in the area, and the environment expected to be present, to determine whether detection is adequate. This analysis is essentially specific and unique to the area being reviewed. In order to do this, they will have to rely on an accurate representation of the coverage blockages within the area, with a reduction in the number of variables (i.e. removal of an algorithm which predicts detection likelihood which may not account for those previously discussed factors).

Having reviewed how a design engineer can use performance-based design to assess flame detector optimisation, we will now review strategies which are often seen in practice but would arguably be best avoided.

Device Layout Based on Achieving a Target Percentage of Coverage

Where a simple percentage coverage is stated as a basic acceptance criterion, this can often fail to consider whether such a basic and generic metric is acceptable in addressing the hazards. For example, an area which achieves 65% coverage may be a 'safer' system than another area which achieves 90% coverage. The driving reason behind this is the specific nature of the gaps in coverage, and the hazards presented in such areas. Take the 65% coverage area. If the gaps in coverage are an accumulation of small negligible gaps, and the hazard is of a reasonably low level, this can be deemed adequate. Conversely, the area which achieves 90% coverage may have a single blind spot of 10% which happens to be the location of the greatest hazard (1, 31).

Even where the acceptance criteria are a simple target percentage, it is pertinent that the designer verifies the coverage graphically. Based on the designer's understanding of the hazards, overall risk, congestion, environment, and technology applied, an assessment of adequacy may deem a design which meets the target percentage unacceptable, or a design which does not meet the target, acceptable.

Alter the Performance Targets to Suit an Existing Design

When using target fire sizes as the metric of acceptability, increasing the target fire size will reduce the number of devices required. Reducing the target fire size will increase the number of flame detectors required. Where financial pressures exist, there is therefore a temptation to increase the acceptable target fire size to reduce the required number of detectors. It is important to remember, however, that flame detectors are designed to detect small fires to act as a mitigation against the fires becoming large. The detectors are also generally designed to detect small fires such as the FM3260 test fire (7), which is a reasonably small fire as previously discussed. If a design, therefore, uses a large fire as the performance target (e.g. >500 kW RHO), it is not possible to verify the design is acceptable, as the test data do not reflect capability to such a target.

Flame detectors designed and manufactured to detect small fires (~40 kW RHO) may struggle to detect excessively large fires due to sensor saturation, for example. If on board algorithms are required to analyse a reading from a sensor, and this sensor is saturated to its maximum reading, analysis of the signal may not be possible. If a detector simply programs a maximum sensor reading to generate an alarm, this may increase the false alarm frequency from the detectors, which is also not desired.

Considering the factors discussed in this chapter, designing to detect a catastrophic type fire such as that shown in Figure 5.15 (33, 34) (left-hand side) is therefore not optimal. The known performance of flame detectors to fires such as that shown on the right-hand side of Figure 5.15 is therefore a more reliable target to design to, with multiples of this RHO allowing performance-based design.

Designing a detection layout to detect an expected fire from, for example, a high-pressure gas compressor would therefore mean designing the system to detect a fire which is generally not tested. How, therefore, can adequate validation of the design occur? In short, flame detection mapping is best served by aiming to detect fires from adjacent processes or third party

Figure 5.15 Worst case fire (top) vs flame detection design fire (bottom).

sources which would cause a failure of the high-pressure gas compressor if it began to impinge on it. Such a design should therefore also reasonably assume that in the event of a spontaneous catastrophic failure of the high-pressure gas compressor, the design would almost certainly detect such a fire. This is an assumption which would of course need to be verified by the designer for the specific application, however.

Note

1. Competence is not an easy factor to qualify, particularly in a niche field like F&G Mapping. This brings in the critical point of *relevant* competence. You may have, for example, a Computational Fluid Dynamics (CFD) expert who has 30 years of experience in the field. Their CFD modelling may even have been applied in the safety industry in gas migration studies, for example. This would not, however, qualify that individual to design an F&G system. This is also an issue experienced in Fire Safety Engineering and is expertly evaluated by Michael Woodrow, Luke Bisby, and Jose L. Torero of Edinburgh University (6. Woodrow M, Bisby L, Torero JL. A nascent educational framework for fire safety engineering. Fire Safety Journal. 2013;58:180–94.) and certainly applies to safety design in the process sector.

References

1. BSI. BS60080 Explosive and Toxic Atmospheres: Hazard Detection Mapping—Guidance on the Placement of Permanently Installed Flame and Gas Detection Devices Using Software Tools and Other Techniques. BSI Standards Limited; 2020.
2. The Offshore Installations. Prevention of Fire and Explosion, and Emergency Response. Regulation; 1995.
3. ISA. ISA TR84.00.07 Guidance on the Evaluation of Fire, Combustible Gas and Toxic Gas System Effectiveness. International Society of Automation; 2018.
4. IEC. IEC 61508 Functional Safety of Electrical/Electronic/Programmable Electronic Safety-related Systems. Geneva, Switzerland: IEC; 2010.
5. IEC. IEC 61511 Functional Safety—Safety Instrumented Systems for the Process Industry Sector. Geneva, Switzerland: IEC; 2017.
6. Woodrow M, Bisby L, Torero JL. A nascent educational framework for fire safety engineering. Fire Safety Journal. 2013;58:180–94.
7. FM A. FM 3260 American National Standard for Radiant Energy-Sensing Fire Detectors for Automatic Fire Alarm Signalling. FM Approvals LLC; 2004.
8. BSI. BS EN 54–10:2002 Fire Detection and Fire Alarm Systems. Flame Detectors. Point Detectors. BSI; 2002.
9. Sizeland E, McNay J, editors. Optimising flame detection layouts in external hydrocarbon processing area: Beyond the theoretical. In: Presented at FABIG Technical Meeting 099. Aberdeen and London; 2019.
10. McNay J. Desensitisation of optical based flame detection in harsh offshore environments. International Fire Professional. 2014;(9).
11. Gottuk D, Dinaburg J. Video Image Detection and Optical Flame Detection for Industrial Applications. Fire Technology. 2012;49(2).

12. Law A, Bisby L. The rise and rise of fire resistance. Fire Safety Journal. 2020;116:103188.

13. Gernay T. Fire resistance and burnout resistance of timber columns. Fire Safety Journal. 2021;122:103350.

14. Duncan G. Flame detector selection—Which one? International Fire Protection Magazine. 2018:72–4.

15. FMGlobal. Property Loss Prevention Data Sheets 5–48: Automatic Fire Detection. Factory Mutual Insurance Company; 2011.

16. Honeywell. Installation Guide and Operating Manual FSX Fire and Flame Detectors Model FS24X. Available from: https://prod-edam.honeywell.com/content/dam/honeywell-edam/sps/his/en-us/products/gas-and-flame-detection/documents/fs24x_operating_manual.pdf.

17. Draeger. Flame 5000 Visual Flame Detector Instructions for Use. Available from: http://www.keison.co.uk/products/drager/Flame5000Manual.pdf.

18. SFPE. SFPE Handbook of Fire Protection Engineering. 3rd ed. National Fire Protection Association; 2002.

19. Koseki H, Yumoto T. Air entrainment and thermal radiation from heptane pool fires. Fire Technology. 1988;24:33–47.

20. Milne D, Davidson I. Flame Detection Assessment Gas Detection Assessment (FDAGDA). Micropack.

21. OGJ. Fire, Gas Mapping Improves Safety, Lowers Cost; 2001. Available from: www.ogj.com/home/article/17220876/fire-gas-mapping-improves-safety-lowers-cost.

22. Milne D, McNay J. HazMap3D, 3D F&G Mapping Software. Micropack (Engineering) Ltd.; 2016.

23. HazDet. Fire and Gas Detection Mapping Software. Available from: http://www.hazdet.com/.

24. MES. AMNIS 3D Fire and Gas Mapping Software. Available from: http://www.mes-international.com/amnis.php.

25. BSI. PD7974–3:2019 Application of Fire Safety Engineering Principles to the Design of Buildings, Part 3: Structural Response to Fire and Fire Spread Beyond the Enclosure of Origin (Sub-system 3). BSI Standards Limited; 2019.

26. Zhen T, Klise KA, Cunningham S, Marszal E, Laird CD. A mathematical programming approach for the optimal placement of flame detectors in petrochemical facilities. Process Safety and Environmental Protection. 2019;132:47–58.

27. Heynes O, Kanno S, Nitta K, editors. Visibility of partially obstructed flames with application to fire mapping. In: Presented at FABIG Technical Meeting 099. Aberdeen and London; 2019.

28. BP. GP 30–85 Fire and Gas Detection Group Practice. Available from: https://pdfslide.net/documents/gp-30-85-fire-and-gas-detection.html?page=1.

29. Monitors G. FL4000H MSIR Flame Detector Performance Report. Available from: https://gb.msasafety.com/Fixed-Gas-%26-Flame-Detection/Flame-Detectors/FL4000H-Multi-spectrum-IR-Flame-Detector/pn/FL4000H-1-0-1-3-1-1-1.

30. Det-tronics. X3301 Multi Spectrum Flame Detector Specification Data. Available from: https://www.det-tronics.com/products/x3301-multispectrum-infrared-flame-detector.

31. McNay J. The Role of Engineering Judgement in Fire and Gas (F&G) Mapping. International Society Automation; 2017. Available from: www.isa.org/Safety-and-Security-Division/FG-June_2017/.

32. BSI. BS7974: 2019 Application of Fire Safety Engineering Principles to the Design of Buildings—Code of Practice. BSI Standards Ltd.; 2019.
33. McNay J, Duncan G, Davidson I, Knox M, Dysart C. Micropack Fire and Gas Detection Design and Technology Course. Micropack (Engineering) Ltd.; 2020.
34. McNay J. Competency in F&G Mapping. International Fire Protection Magazine. 2019.

6   Gas Detection Technologies

As with flame detection, understanding how gas detection technologies operate and applying this knowledge in design are critical.

Consider this scenario. IR gas detection has been installed in an analyser house where the hazard in question is a hydrogen leak. As this chapter will expand upon, IR-based gas detection will not detect the presence of hydrogen, as hydrogen does not absorb the IR beam, and therefore does not provide a response. When the F&G Engineer queries the application of IR detection to detect a hydrogen risk, they are instructed that the devices had been specified without considering this. Once installed, however, and after being told that they are serving no purpose, the belief that 'well IR is better than nothing, leave them in' prevails.

The coupling of misunderstanding the capability of detection technology, and reluctance to be perceived as 'reducing safety' by removing a detector, leads to a waste of money on maintenance of devices which provide no safety benefit whatsoever, leaving the analyser house with no detection capability of the hazard in question.

This scenario highlights not only the importance of consideration of the specific hazards during design and selection of the appropriate technology, but also the importance of management of change where errors of the past can be rectified to ensure that the systems are adequate through its lifecycle. In many cases, it may not even be an error but simply a design decision which is no longer relevant for the site. Leaving such technologies and devices which are no longer useful creates a systemic route to failure as maintenance routines become burdensome, ultimately leading to the neglect of a safety system which *is* required.

Over the last 30 years, gas detection technologies have not changed drastically. The operating principles of the devices are fairly well understood, particularly when it comes to IR-based point- and open-path-based devices. While emerging technologies like laser-based gas detection, camera-based gas detection, and the more established ultrasonic-based gas detection are useful in specific applications, this chapter will focus closely on the traditional flammable gas detection technologies.

At the time of writing, infrared camera-based gas detection technology is emerging. Such a technology appears to successfully detect leaks at impressive

DOI: 10.1201/9781003246725-6

distances, potentially removing the need for multiple gas detectors placed within the facility. While the technology appears impressive, this does, however, present some issues in design. Examples include areas where high congestion exists, and the gas can accumulate in a fashion where it is hidden from the camera. Such applications are of high risk, as the gas is accumulating in a congested area (which will be discussed further in Chapter 7); therefore, traditional detection technology would appear to be more optimal. Another issue is the determination of alarm level. It is an important concept that not all gas releases present a hazard. Typically, designs aim to detect a gas cloud of sufficient volume which would present a risk of explosion. When using traditional detection, this approach is proven in use and simple to design. When using camera-based technologies, the application of reasonable set points for alarm raises some challenges which require further research. For these reasons, this emerging technology will not be included in this book.

Catalytic

The prevalent gas detection technology on older installations is that of catalytic-based gas detection. While there are some limitations associated with this technology which will be discussed in greater depth through this chapter, the one major concern, regarding their application in harsh environments, relates to its mechanical construction.

As will be shown, the detection elements are contained within a protected housing, where gas must pass through a sinter. Such sinters, or sintered disks, are prone to blockage, and can prevent gas from being able to pass freely into the sensing chamber.

The catalytic-based gas detection technology uses catalytic bead detectors which require gas to be 'burned' within the detector. This burning gas then produces a gas reading which corresponds to LEL/LFL readings.

Well-documented limitations associated with the technology include the poisoning of the catalyst and blocking of the sintered disks (both of which can lead to unrevealed failures). Where the device is poisoned, it is credible that no alarm will be generated. The devices also generally cannot be used in an inert atmosphere. Sensors can also 'drift' and require regular calibration. Exposure to high concentrations of gas can also damage the sensor and impair future performance. Gas-rich environments may also saturate the device, meaning that an alarm is not generated.

Figure 6.1 represents the operating principle of a typical catalytic gas detector.

The response times of catalytic gas detectors are also comparatively low compared to other more modern devices. Response times can be in the range of up to 30 seconds; however, specific detector manuals should be consulted to verify the specific detector response time.

The devices are also generally highly consumptive. The heating element within the device increases the cost of operation.

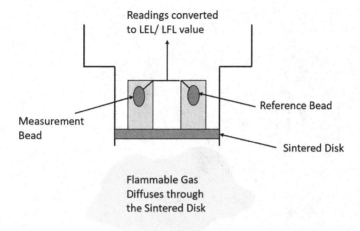

Figure 6.1 Operating principle of catalytic gas detection.

Due to these issues, certain regions and operators do not recommend their use for general application hydrocarbon gas detection.

They can, however, be used in applications where modern IR solutions are not applicable. There exist gases which do not absorb IR at the required wavelength and therefore cannot use IR-based detection.

Take the example from the scenario at the beginning of this chapter. Hydrogen gas cannot be detected through mainstream IR means. Catalytic gas detectors can, however, detect hydrogen gas, along with virtually all gases which are flammable. This is a result of the operating principle which 'burns' the gas, as it permeates into the device.

This advantage of being able to detect virtually all flammable gases is not the only advantage. The technology is well known and very well established, having been developed by Dr Oliver Johnson in the mid-1920s. The requirement behind the technology development was to prevent explosions in storage tanks and gasoline tankers. The devices are also of a reasonably low cost by comparison to more modern technologies; however, this may be offset by a higher maintenance cost as a result of the aforementioned limitations.

It is important to note that this discussion presents the technology in general terms. Specific device manufacturers should be consulted regarding their devices and how they combat the limitations and exploit the advantages of the technology.

Figure 6.2 shows a deconstructed catalytic gas detector.

SINTER*

Figure 6.2 Deconstructed catalytic detector.

Infrared Point Gas Detector (IRPGD)

As a result of the limitations of catalytic-based gas detection, IR-based gas detection has risen in prevalence. While these devices are still susceptible to blockage where a weather shield is used, the likelihood of blockage is not as high. IR gas detection has risen in prevalence and is now the preferred flammable gas detection technology by many oil and gas operators, in many regions across the world. The technology is fairly well established, with the first units being supplied into the North Sea as early as 1985 (1).

Figure 6.3 represents the operating principle of a point IR gas detector. While individual devices will be constructed in their own unique way, the operating principle remains reasonably consistent. The detector housing, which includes the processing equipment and electronics, is connected to an IR source which projects the IR from the device, which is reflected back into the device. On a point gas detector, this distance is relatively short (generally <150 mm). Where gas passes through the beam, the signal drops and is proportional to the %LEL/LFL. A reading is then generated on the basis of this degradation of signal.

In order to protect the optics from the environmental elements, weather shields can be placed on the devices. These also protect against other environmental obscurants such as dirt and foreign bodies including insects/bugs, although these can still find their way inside. The shields have gaps which allow the gas to freely flow across the optics. In extreme circumstances, these can become blocked. An adequate maintenance routine is therefore important in ensuring adequate detection. Also, in certain clean environments, the devices can operate without such a weather shield, in a 'naked optics' form.

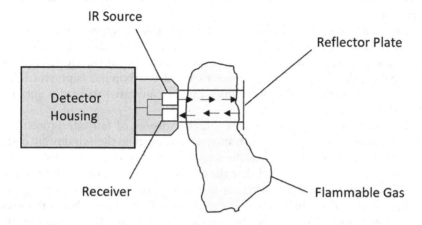

Figure 6.3 Operating principles of point IR detectors.

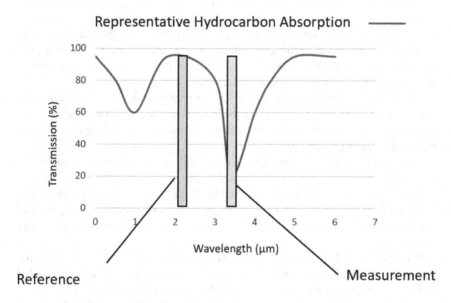

Figure 6.4 Operating principles of IR detectors.

Different gases will have a varying transmission behaviour across the electromagnetic spectrum. Devices are therefore calibrated to a specific gas, or range of gases, which must be carefully selected on the basis of the anticipated gases on site.

Figure 6.4 shows an example transmission footprint of a fictional hydrocarbon gas. The figure shows the wavelengths in which a signal will freely pass through the gas, and those wavelengths at which an IR beam will be absorbed.

The figure shows a suitable wavelength for both reference and measurement signals. As can be deduced, the reference wavelength will freely pass through the gas, and when the signal of the measurement wavelength begins to weaken, this will present evidence that hydrocarbon gas is present. This ratio change between the two values can be calibrated to read a specific %LEL/LFL for specific gases.

IR-based gas detection provides a higher degree of fail-safe protection, through application of optical testing to ensure that the window is clear and that gas should pass across the window it will be detected. Anecdotally, some operators have claimed that the current generation of IR point-based gas detection can be installed in an 'install and forget' manner. No maintenance is carried out, believing that the device will tell them when it requires attention. While there may not be consensus on whether this is appropriate when dealing with optical safety devices, it demonstrates the confidence the industry has on the reliability of the technology.

IR-based detection also benefits from a comparatively faster response time in the presence of gas to a catalytic-based detector, and generally consumes less power. Heating elements to prevent condensation on the lens can, however, increase the power consumption of an IR-based device. Despite this, the power efficiency of the technology is demonstrated by some wireless devices which can have a battery life up to 2 years.

When considering that the limitations of catalytic devices becoming poisoned, this is not an issue with IR-based devices as a result of their basic optical nature. This also means that they will provide a reading even in gas-rich and inert environments.

Despite these benefits, the technology is still susceptible to limitations. As previously discussed, IR-based technology is unable to detect certain specialised gases such as hydrogen. The technology also cannot detect toxic gas hazards; however, they can be used to infer the presence of toxic gas, and this toxic gas should be entrained in a flammable gas which the IR detectors can detect.

An additional limitation, as previously mentioned, is the potential for the weather shield to become blocked, which would present an unrevealed failure. Making matter worse, some detectors are supplied with a fine filter to exclude dust or moisture. These filters can become blocked, preventing or delaying detection of a hazard. Such failures, again, may not be revealed.

The specific filter selection of an IR detector can impact how sensitive the devices are to target gases. Where multiple gases are present, for example, IR detectors may be more sensitive to certain stream compositions. Once again, consultation with the detector manufacturer is critical to ensure that the correct device is selected on the basis of the hazards expected. Care must be taken to ensure that an underreading of the flammable gas is not likely.

With respect to further limitations of the technology, the initial purchase price for an IR-based device is likely to be more expensive than a catalytic-based device.

While filter selection and technology are crucial in successful detection, it is not something which typically appears in or influences the graphical mapping portion of a review. The process of mapping is discussed in greater detail in Chapter 7, but the selection of appropriate technology is just as, if not arguably more, critical than the modelling phase.

Once again, it is important to discuss the specifics of a device and the application with the manufacturer.

Infrared Open Path Gas Detector (OPGD)

Developing on from the point-based IR gas detector was the development of the IR open path gas detector. These devices have risen to prevalence and continue to provide an attractive solution as a result of their ability to span large distances and reduce detector counts.

The development of the technology was not without issues, however. Early OPGDs were known to be susceptible to the harsh environment of a typical petrochemical application. Issues associated with alignment, stability, and reliability were exacerbated by regular false alarms from environmental stimuli like water mist/fog (1).

Development of the technology in the last 20 years has seen drastic improvements in their stability, reliability, and false alarm resilience. The application of a Xenon flash lamp and the improved filter selection have seen that these devices become the detection technology of choice for many facility operators. The technology is now generally well received in the hazardous industries (2).

Figures 6.5 and 6.6 show the operating principle of applying the infrared technology in an OPGD.

The OPGD technology presents many benefits over traditional technology. The first and most obvious being the increased ability to detect clouds

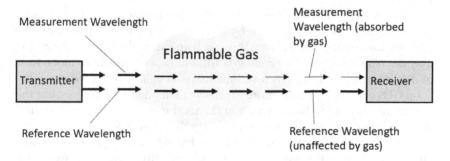

Figure 6.5 Operating principle of OPGD.

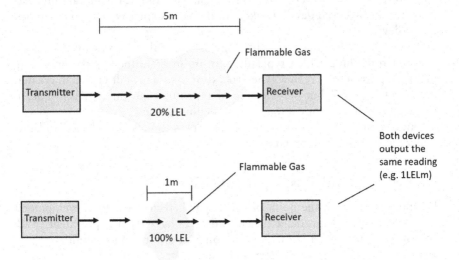

Figure 6.6 OPGD concentration vs volume.

over large areas without an excessively high detector count, with a reasonably fast detection response (compared to catalytic-based devices).

Similar benefits as those afforded by the point IR devices were also carried across such as the inability to be poisoned and their ability to operate in inert environments and detect gas even in gas-rich environments.

The devices are also able to operate in areas of high airflow. While these environments can cause gas clouds to by-pass point detectors, as a result of the open paths greater footprint, it becomes harder for clouds to avoid contact with the beam. The technology also lends itself perfectly to the notion of gas cloud mapping which is the primary form of gas detection mapping. It also addresses the fundamental issue of gas detection detecting clouds which will present an explosion hazard rather than small, localised accumulations of gas (obviously performance target dependant). This will be discussed further in Chapter 7.

The devices are not, however, without their limitations. IR OPGDs are generally limited to the detection of C1-C6 hydrocarbons, and the exposed beam presents some risk of false alarm. With such an exposed sensing element, this may not truly be able to be eliminated without moving to a laser-based form of detection (which brings its own limitations as will be discussed).

OPGDs are also generally more expensive than point type devices, but this is regularly offset by the reduction in the total number of devices required, with savings carried forward in OPEX as well as CAPEX.

An additional limitation to consider is that where an open path spans a large area, an alarm from this device will not inform the operator of the specific location of the gas. Point gas detectors can provide this benefit of a spot flammability reading which OPGDs cannot provide.

Once again, it is important that device manufacturers are consulted on the capabilities of their own devices.

Toxic Gas Detection

Before discussing the most recent development in OPGD technology (which is often used for toxic gas detection), a brief note on traditional toxic gas detection technologies is presented.

Toxic gas detection has generally been provided by point gas detectors of either electrochemical cell or semi-conductor-based technologies.

Electrochemical cell and semi-conductor detectors operate on a similar principle of gas flowing through a permeable membrane and creating a reaction within the detection device (either reacting with an electrolyte or semi-conductor in each respective technology).

Each device suffers from similar limitations (slow response, potential for drift, etc.), but in the absence of being able to apply IR-based technology to the detection of low concentration toxic risks, facilities are living with the limitations.

Deciding which technology to apply between electrochemical or semi-conductor primarily depends on the environment. Factors such as humidity and temperature flux will determine which device is best suited to the application, and guidance on this should always be sought from the detector manufacturer.

Laser-Based Open Path

Moving back to open path technology, the use of tuneable laser diode technology emerged as the next step from IR OPGDs. Despite the increase in cost over traditional OPGDs, facility owners are being persuaded by the technology's ability to ignore environmental stimuli, and subsequently reduced false alarm frequency.

Laser-based devices look for the specific absorption line to a specific gas, rather than a relatively wide band of IR. The absorption line of interest is known by its wavenumber. Tuneable diode lasers minimise cross interference by locking on a single line where there is practically no cross-sensitivity due to the sharpness of the absorption lines (2).

These devices can be highly sensitive and detect very small concentrations of gas which is useful in toxic gas detection. In extreme circumstances (i.e. H_2S > 10,000 ppm in stream), laser open paths may be necessary to aid in detecting such high concentrations quickly and reliably, particularly when considering the limitations previously discussed in electrochemical cell/semi-conductor type devices typically used in detecting toxic gases. When a fast response is required, or exceptionally high-concentration environments are possible, the reliability and response time of a laser-based detector may be preferable.

The benefits of a laser-based open path are therefore clear. The benefit of using an open path in reducing the number of traditional point detectors can reduce maintenance burden and (possibly) the cost of detector procurement. As with IR OPGDs, they are unaffected by high concentrations of gas and inert environments, and operate in high airflow environments. The lasers can also be tuned to specific gases, making their reliability unparalleled with the devices practically immune to cross sensitivity. This also reduces interference to sunlight, water mist, and other false alarm sources.

The response time of the devices is also impressive, also benefitted from the fine-tuned nature of the devices.

The sensitivity can read into the ppm range, making them far more accurate and sensitive than traditional IR OPGDs.

The disadvantages, however, relate to some of the advantages. The fine tuning, for example, of the detection fingerprint means that they generally cannot be relied upon for general process area gas detection, unless the facility is processing a single gas. These devices also required to be calibrated to the specific gas.

The other main limitation is the cost of the devices. Despite reducing the number of traditional detectors which would normally be required, laser-based OPGDs can still be too expensive for many projects. This may, however, change with time, as component costs reduce and manufacturing quantities increase.

Ultrasonic/Acoustic

Ultrasonic, or acoustic, gas detection, as its name implies, applies acoustic-based sensors to listen to an environment and detect changes in sound pressure level (SPL) outside the scope of human hearing. The sensor and electronics monitor ultrasonic frequencies (25 to 100 KHz), while excluding audible frequencies (0 to 25 KHz) (1). The detector listens for the leak rather than measuring the accumulation of a flammable cloud. Ultrasonic noise is generated when gas is released from a high-pressure to a low-pressure area.

The change in SPL travels across the facility; therefore, the release itself (i.e. the flammable gas) does not need to come into contact with the device, as is required with catalytic or IR-based detection of flammable clouds. This provides a benefit with respect to response time. If the device does not have to wait for the physical properties of the release which could cause an explosion to migrate across the facility, detection can be near instantaneous. It does, however, mean that the devices cannot differentiate between flammable and non-flammable releases, and are unable to provide a reading on concentration or volume of the cloud.

The devices are not affected by wind speed/direction, but can be affected by solid boundaries which may impact the transfer of the ultrasonic signature of the release.

The background noise will impact on detection capability. The higher the background ultrasonic noise, the smaller the detection envelope will be. It is important to note that this only refers to background noise in the same frequency, as the detector is operating. An area may, for example, have a high background noise level in the human audible region, but in the ultrasonic region, it is very quiet, meaning that the devices will have a detection envelope unaffected by the background audible noise.

Background ultrasonic noise levels should be mapped in order to adequately design such a system. The application of set points above the background noise level, and time delays are often applied to reduce false alarms from such a system.

Despite the devices solely detecting the 'hiss' of a gas release, the devices have become more robust over time in harsh external environments. With the technology being more reliable, this provides a useful additional layer of protection regarding mitigation of flammable gas releases. The devices are often used as a useful supplementary detection system for high-hazard areas.

Cross voting of IR detection and acoustic detection is often reliably applied, as there is no single false alarm source which would activate both technologies. Therefore, if both an IR gas detector and an acoustic detector are in alarm, it is likely that gas is present.

References

1. Sizeland E, McNay J. Using CFD to Optimise Gas Detection Layouts: Are We Barking Up the Wrong Tree? In: FABIG Technical Meeting 99; 2019. Available from: https://www.fabig.com/publications-and-videos/technical-presentations-videos/technical-meeting-099/.
2. McNay J, Duncan G, Davidson I, Knox M, Dysart C. Micropack Fire and Gas Detection Design and Technology Course. Micropack (Engineering) Ltd.; 2020.

7 Flammable Gas Detection in Process Areas

Within gas detection design, in numerous applications, we suffer from the issue that the most appropriate method of mapping is hard to explain but easy to implement, and the least appropriate method can be easy to explain and challenging to adequately implement.

Take the following scenario: a conversation during a project meeting where a gas processing module is adjacent to a utility area. The volume of gas which would cause an explosion is determined, and implemented as the target—throughout the whole volume (both utility and process areas). The design is queried, as a result of the placement of gas detection in the utility area where there are no sources of gas release. Unfortunately, an invisible wall is not placed between the gas processing area and utility area which would inhibit the gas from moving to the adjacent congested area.

The belief that gas detectors are only required in the area where the gas is processed is a common misconception. It is easy, however, to explain that 'gas detectors are to be placed where gas is processed'. The belief therefore takes hold.

There are similarities in the recent approach of placing detectors based on a limited number of simulations. Easily explained by stating designers should place gas detectors where gas is expected to travel. On the face of it, this makes sense and is therefore commonly believed as an adequate solution on its own. The problems associated with this strategy in isolation are numerous, however. As previously mentioned, it is easy to explain but challenging (and often inappropriate) to implement.

Picture the following all too credible conversation:

F&G MAPPING PROVIDER: 'Gas will go here when released—we've run simulations which show this'

3RD PARTY PEER REVIEWER: 'But gas poses no threat there—there is no congestion or confinement and no risk of escalation'

F&G MAPPING PROVIDER: 'Yes but it is likely to go there.'

3RD PARTY PEER REVIEWER: 'that likelihood is based on 0.0001% of possible scenarios. What about this vast congested area which has no gas detection at all? If gas happens to go there, it will present a significant explosion risk which could take out the plant. Has this been considered?'

DOI: 10.1201/9781003246725-7

F&G MAPPING PROVIDER: 'The gas only goes there in 5% of the simulations we ran so we don't need gas detectors'

3RD PARTY PEER REVIEWER: 'This conclusion is based on 10,000 simulations compared to billions of potential scenarios credible in such an unpredictable environment'

F&G MAPPING PROVIDER: 'That is still 5,000 more scenarios than other software/consultants would have run . . .'

The running of simulations to understand air flow movements and potential gas cloud behaviour is useful and certainly has its place in certain applications of performance-based F&G mapping. Without looking at the challenge of gas detection placement from a fundamental perspective can lead to a perilous pursuit which misses the intent of flammable gas detection design. This chapter will aim to present the challenge and provide guidance on how to select the correct approach based on the facility. The main outcome the author hopes that this chapter will achieve is to present that each application and subsequent risk profile should be reviewed before deciding which approach is best placed to optimise the detection layout. Sometimes a prescriptive approach is best placed, sometimes a volumetric or geographic approach is best, and sometimes a scenario-based approach is best. In some cases, a mixture of each is suitable.

It is simple to understand the notion of placing detectors where the gas is expected to go, but it doesn't mean that this is the most appropriate solution in all cases. In the 17th century, people would look across the land believing the world to be flat. When the claims of Galileo did not compute with what was easily explained, he was charged with heresy. While tenuous, there is a link between those who have certainty that a specific methodology is always the best route to follow. The late Christopher Hitchens put it best when he said, 'since we have to live with uncertainty, only those who are certain leave the room before the discussion can become adult' (1).

Application of Flammable Gas Detection

Within the processing of flammable materials, the risk of a break in containment is inherent. The release of such material can present a multitude of hazards including the obvious risk of explosion and the potential for fire, but such releases can present the risk of exposure to toxic gas, for example. The main focus of this chapter is on the flammability risk associated with a break in containment.

The first and main point with respect to flammable gas is that it is currently not feasible to detect all leaks. The focus, therefore, is to determine which releases will be of sufficient concern to create a potential hazard. Such releases become the target for our fixed flammable gas detection system.

When considering detector placement, flammable gas detectors were traditionally placed adjacent to leak sources, or based on the buoyancy of the gas (i.e. detectors in an area with a Methane hazard would be placed at high elevation as Methane is lighter than air). This practice was soon found to be inadequate; however, as it became clear that the behaviour of a gas cloud is determined by the characteristics of the release and the environmental influences in the area of release. With the slightest increase in stream pressure, the placement of detectors adjacent to the release point can quickly become ineffective and unreliable.

Gas Detection Performance Targets

In the wake of the Piper Alpha disaster, it was clear that a significant body of work and academic understanding of gas cloud behaviour and explosions analysis in congested and confined hazardous areas existed but was generally kept within those practicing F&G analysis within operators. With such information and knowledge not being shared on a wider scale, this created a systemic failure with respect to continual development and deployment of fire and explosion safety. After Piper Alpha, this changed, creating an open and transparent safety culture.

In the discussions around improvement of safety systems such as the fixed flammable gas detection system, one must fundamentally understand the philosophy behind the application of the technology. The primary objective of flammable gas detection is to detect the presence of gas clouds, which have accumulated/are accumulating to the point that they present a risk of explosion overpressure if ignited. This informs the traditional and primary methodology of gas detection design, the target gas cloud (TGC) method.

The TGC method aims to define the cloud which would present an explosion overpressure, then set a target gas cloud for detection below this critical threshold. The critical cloud size has been shown to be heavily influenced by the congestion and confinement within the area of concern.

The TGC approach is based upon the findings and literature presented in HSE OTO 93–002 (2).

This document presents data on the overpressures associated with ignition of various gas accumulations in various congested applications. The primary, and most referenced, finding from the report is the determination that in an area with blockage ratio between 0.3 and 0.4, a 6 m cloud of stoichiometrically mixed flammable gas will not result in an explosion overpressure of 150 mBar (flame speeds over 100 m/s or 125 m/s, respectively) or more. The calculation of blockage ratio will shortly be discussed; however, in this instance, this represents the ratio of the volume which is congested/confined. The importance of the 150 mBar explosion overpressure is that this value is widely accepted as the minimum threshold for pressure-induced damage.

When considering increasing blockage ratio, the cloud size which would result in damaging overpressures reduces. As the blockage ratio reduced, the cloud which would be required to cause damage of concern increases.

It is important to note, with this varying degree of blockage ratio and the specific impact on target gas cloud, is an area where further research would be highly beneficial. Various rules of thumb have traditionally been applied (as will be discussed later in this chapter), but detailed research to determine cloud sizes versus blockage ratio would be of great benefit.

In calculating the blockage ratio, the following simple calculation can be applied:

- Create the boundary zone and treat this as a cube
- Each side contributes ~17%, or ~0.17, to the blockage ratio
- If all six sides are solid, blockage ratio = 100%, or 1.0
- Example blockages:

  - Solid Deck, open on all sides—blockage ratio = ~0.17
  - Solid Deck, solid West side, open on all other sides—blockage ratio = ~0.34
  - Solid Deck, solid West side, open on all other horizontal sides, grated ceiling—blockage ratio = ~0.34 + a reasonable value for grating*

*Grated floors/ceiling and congested areas which are not fully solid boundaries require engineering judgement to determine blockage ratio. This value can be between 0.0 (open) and 0.17 (solid).

Figure 7.1 presents an example zone with two adjacent areas presenting differing blockage ratios.

The approach generated from the HSE OTO was more recently indirectly validated by a study by the Institute of Chemical Engineers (IChemE) (3). This was in light of developments in analysing gas cloud behaviour and provides a more recent reference for the effectiveness of the approach in light of the dated nature of the 1993 HSE report.

The finding of the HSE OTO report around specifying a target cloud is often referred to as the 5 m spacing rule. While this is an oversimplification

Figure 7.1 Graphical representation of blockage ratio.

of the true nature of the findings, it was still a major step forward in gas detection design. Despite this improvement, gaps do exist in the findings of the OTO.

Certain aspects of gas detection were not addressed in the OTO, such as gas cloud and explosion behaviour in open areas with little or no congestion. It also did not address aspects like HVAC gas detection (this will be discussed in Chapter 8). Migration detection/perimeter-based gas detection design was also not addressed in the OTO.

Such gaps were generally addressed in operator-specific guidance, practicing third-party consultancies, and more recently in the ISA and BSI guidance on F&G Mapping.

In particular, the omission of explosion analysis in open-based areas required to be addressed by practitioners and operators, as this form of application is common on most facilities. For example, top decks on offshore facilities, and onshore process areas can present large open spaces with hydrocarbon processing. Such areas do not present the blockage ratios addressed in the findings of the OTO.

Essentially, operators began to specify reasonable target gas clouds for such areas, and this is where the 10 m cloud and dilute cloud factor of 3 emerged (to be discussed later).

The resulting TGC method provided the philosophy of fixed flammable gas detection, but the data to support the various target cloud sizes required further research to reach true optimisation of fixed detection designs.

Most notably, the emergence of CFD modelling of explosion hazards may provide the answer to the optimisation of the TGC method. While effort is being placed on CFD for dispersion analysis, more useful application of the tools may be found in analysing critical cloud sizes and their effect on explosion overpressure.

Despite this research which can help a designer in making decisions, as with flame detection, a performance target is still the most critical aspect of a gas detection design, and the application of targets, acceptability criteria, and modelling outputs are surrounded by uncertainty.

Performance targets require to be verifiable. As with flame detection, any modelling of coverage should address the target specifically, and should also be capable of representing the available technologies adequately. Only then can the designer make a reasoned decision on whether adequacy has been achieved.

Once the designer has reviewed that the risk profile and gas hazards present, along with the nature of the facility, the decision on methodology is made which is critical in influencing what a performance target is. For example, where fixed gas detection is to be used to mitigate an explosion hazard in a congested area, the volumetric approach will likely be specified. The target, therefore, becomes a cloud size and a required percentage of the volume which needs to be covered (this target percentage approach suffers from the same issue discussed when reviewing flame detection). Conversely, where a gas sampling approach is being used, a scenario-based approach may be applied. The target here would switch to a target number of scenarios which required to be detected. For example, in a simulation run of 20,000 scenarios, detectors must be successful in 95% of simulations. When applying a design based upon such scenario-based dispersion analysis, a controlled environment is typically desired. In such an environment, scenario-based mapping presents a useful solution, as the uncertainty around environmental influences can be reduced (4).

The strengths and limitations of each approach are well documented in the BSI guidance on hazard detection mapping (5).

Gas Detection Mapping

Where the performance targets, risk profile, and proposed methodology require it, gas mapping software is often applied to assist the designer. The tools will typically provide a 3D representation of the area to be assessed, with coverage analysis shown by 3D iso-volumes or 2D slices through the 3D environment.

The original hazard presented in the HSE OTO report (6 m cloud) has traditionally been represented by a hard-edged sphere of the relevant diameter. Initial programs represented this 3D geometry in only 2D slices (6). It is clear for all that such a hard-edged shape of a gas cloud was not credible in reality (although is possible in certain situations). It is arguable, however, that the hard edge sphere is just as credible considering the almost infinite number of variables, as any individual simulation. As a result of this, and the perception that a perfectly stoichiometrically mixed sphere of gas is the worst-case scenario for explosion potential, it is still widely applied today.

There has been some work on the nature of gas dispersion behaviour. In particular, a JIP was presented which investigated cloud behaviour in congested areas (7). This analysis looked at dispersion behaviour with a 3D array of detectors which helped to understand the variance in dispersion from the highest concentration to the air-rich environments.

As would be expected, the JIP demonstrated that for a typical gas cloud, the concentration of gas decreased as distance increased from the high-concentration central point (~200% LEL). This reduction from 200% in the centre to 0% at the extreme is also a valuable input to the TGC method, allowing detector set points to be considered in the analysis.

Nominal values were then applied to the OTO 5 m cloud, where the central high-concentration cloud of 5 m was assumed to be 60% LEL throughout, with a surrounding 15 m cloud of 20% LEL, accounting for this gradual drop off in concentration from the centre of the cloud to the extremities. Note here that 60% LEL and 20% LEL were used in correspondence with traditional High and Low set points for gas detection.

While considering the central dense cloud as 60% LEL and the outer dilute cloud as 20% LEL is not conservative from an explosion potential point of view, it represents a conservative approach from a detection modelling point of view. If the cloud in reality is higher than 60% LEL, the reliability of detection increases. It is therefore a worst-case scenario from a detection perspective to assume the low level of gas concentration when modelling gas detection coverage.

The JIP also allowed OPGDs to be modelled more realistically, by modelling the accumulation of gas across the beam path, which will heavily influence successful detection (as discussed in Chapter 6).

While this approach of including gas dispersion behaviour in a volumetric design is useful, there is a caveat of conservatism required. The method allows designers to consider the dispersion of a cloud from high concentration, gradually out to zero gas readings. As detector set points move lower and lower, this could allow designers to set low set points for both alarm and control action (i.e. 10% LEL), and end up modelling sufficient coverage of gas clouds of an excessive diameter, which may not occur in reality. As with any design methodology, if an area is congested or confined and the design results in significant gaps in coverage (i.e. large spaces between detectors), this should be reviewed carefully, as a cloud of gas in this area *will* remain undetected. The facility operator must therefore decide if the risk of this is acceptable, regardless of whether the mapping report shows theoretical 'good coverage'.

Figure 7.2 shows the application of two OPGDs in an offshore process module and the resultant coverage afforded by them.

The previously mentioned benefit of the JIP was the incorporation of an approach which allows OPGDs to be more accurately modelled. Often OPGDs will be modelled to show coverage as a basic 'sausage' shape. This would assume that if a cloud simply contacts the beam, successful detection will occur. As discussed in Chapter 6, OPGDs require a certain concentration of gas to cross a certain amount of the beam in order to generate an alarm. This basic sausage of coverage is therefore not appropriate.

Where volumetric (or geographic) modelling is presented, reviewers will often see this cylindrical shape of coverage. This representation of coverage is appropriate for point type detection, as this will generate an alarm if the cloud contacts the device at an acceptably high concentration, but it is not acceptable for OPGDs. Mapping therefore should use the following in calculating 'beam attenuation' through the flammable gas, to give a more accurate reading of detector coverage.

Figure 7.3 shows a basic representation of how beam attenuation affects the alarm thresholds of OPGDs.

Hazard Representation

While we know that a gas cloud is unlikely to have a solid edge, or perfectly linear decreasing concentration from the central core, we also do not know

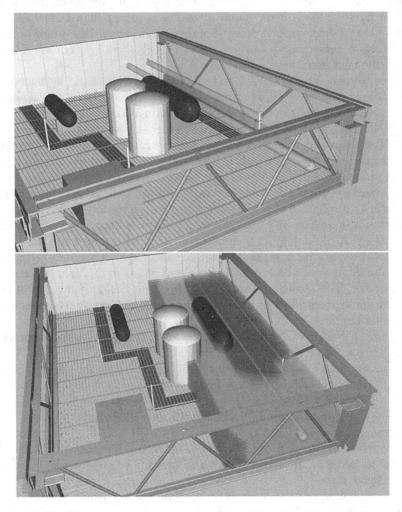

Figure 7.2 Typical simple 3D gas detection assessment (beam attenuation).

what the cloud will look like on any given day, regardless of the number of simulations we can run.

When considering the probabilistic nature of design of active safety systems, anecdotally it has been noted that retrospective simulation studies in post-incident analysis even struggle to reflect the reality of what occurred. This is even when the simulation engineer knows the exact conditions which occurred on the day of release. If the current generation of simulations struggle to reflect gas cloud behaviour when we know the conditions of the day, this should be considered when we use these tools to design a safety

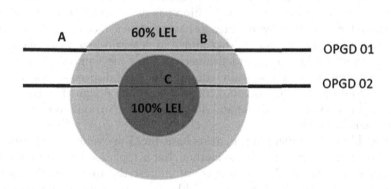

Figure 7.3 Beam path absorption principle (beam detectors OPGD01 & 02 with different readings provided by concentrations A, B & C).

system which relies on their assumptions, when we don't know the conditions which will be present on any given day of release.

If we are to address the critique of the TGC method that hard edge spheres do not occur in reality, then the scenario-based approach would potentially have to place detectors throughout the volume to detect such a vast array of unique cloud shapes and sizes.

This is where the analysis of the facility and the best placed methodology is critical, and in some instances, a combination of the TGC method with a collection of simulations to verify the extent of the hazard may be a beneficial solution.

Regarding the placement of resources in researching further optimisation, the industry would seem to be better served focusing on developing the TGC method to add further inputs around various critical cloud volumes with varying congestion and confinement values.

Target Gas Cloud Method Applying Dense and Dilute Clouds

Regarding the aforementioned JIP on gas cloud behaviour, this was initiated as part of a wider study on gas cloud behaviour in process modules but was particularly useful in the field of gas detection design. The data produced were reviewed by Micropack (Engineering) Ltd. along with oil and gas operators to allow the data to be used for the improved effectiveness of gas detector placement. This work, along with the HSE OTO, resulted in the TGC method previously discussed.

The benefit of the TGC approach allowed F&G Engineers to review the blockage ratio of a volume and assign a target cloud size as per the HSE OTO, then subsequently assign concentration characteristics of the cloud to optimise the placement of devices.

As discussed, this did provide a limitation in that designers could, dependent on the set points assigned, position gas detectors 20 m apart and the methodology would show the area as having sufficient coverage (essentially creating the same critique as that aimed at scenario-based gas mapping). This danger pushed some practitioners to limit the number of variables, which would allow optimisation without additional dispersion assumptions which could create the potential for inappropriate design. The result was the application of fixed dense and dilute clouds for various congestion and confinement factors.

The dilute cloud method uses as its basis the '5 m cloud', for example, as described in the target gas cloud method, but a applies a dilution factor. For example, if 60% and 20% LEL were of interest, the approximation would be a 5 m diameter cloud with 60% LEL edge (increasing in concentration to the centre of the cloud in reality), and a 15 m diameter cloud with 20% LEL edge (with concentration reducing from 60% at 2.5 m radius from the centre to 20% at 7.5 m radius from the centre of the gas volume). Many practitioners of this method, however, do not allow alteration of the set points to allow the increase/decrease of the detectable gas cloud, and it is locked at 60% LEL representing the dense cloud, and 20% LEL representing the dilute cloud. The values of the target gas cloud (as described earlier) are taken, and a dilute cloud factor of 3 is applied to represent the lower set point, as was found in the JIP analysis. This then provides a simple assessment of dense and dilute clouds, but removes the ability of the assessment to take account of large, low concentration, clouds which provide an optimistic representation of gas detection coverage.

While clouds in reality will not be spherical clouds with a fixed diameter, this approach is widely regarded as a suitable performance-based approach to applying the OTO findings, without allowing an overly optimistic view of gas cloud behaviour and coverage. It should also be noted that there is no data to suggest the dilute cloud factor of 3 is appropriate beyond the 5 m dense cloud; however, in the absence of further research/data, many operators have applied this value to all target gas clouds.

Has the TGC Approach Been Applied?

With the development of emerging gas detection placement methodologies like scenario-based gas detection design, the question emerges of whether the traditional TGC methodology (2) has actually been applied in practice.

When an industry has not suffered a disaster on the scale of Piper Alpha for a prolonged period, the increased pressure to optimise and reduce costs is an inevitable systemic creep towards failure. The pressure to reduce gas detector numbers would be of particular concern when we consider the historically poor leak detection rates in the North Sea, for example.

The combination of poor leak detection rates and a drive towards optimisation push the industry towards innovative techniques in detector placement.

It is of paramount importance therefore that those developing new methodologies fully understand why the gas detection rates are 'poor', and what the fundamental underpinning of gas detection design entails. We must also understand the true nature of what is installed in the field, without making the assumption that the recommendations from documents like the HSE OTO are applied in practice.

In reference to this, there is an assumption that the TGC or 5 m spacing approach has been applied on existing offshore installations and this therefore must be blamed for the poor leak detection rates. Industry must consider if this is true. If that assumption turns out to be false, it could lead to the development of methods which are fundamentally flawed which could have catastrophic consequences

Hilditch et al. (8) demonstrated that from a sample of 27 detector layouts that a TGC arrangement occurred in only 44% of cases, while a leak detection approach (a form of scenario-based approach) was present in 66%. There was an overlap in five cases where both TGC and leak approaches are applied. It was also demonstrated that those areas where the TGC approach was applied presented a far greater average coverage with fewer detectors than the areas monitored by a leak-based layout.

This inconsistency in approach makes it difficult to determine which methodology is causing the lack of detection reliability (if the methodology is in fact the cause of poor detection rates, which also remains unclear). It also creates the credible position that the historically low gas detection leak rates in the UK North Sea (8–10) will be further exacerbated by the application of a 'new' approach (i.e. scenario-based design) which may already be in place and failing in its application.

While the consideration around poor leak detection rates inevitably focuses around the methodology of detector placement, of greater importance is the question of whether the leaks would be expected to be detected at all. As the fundamental application of gas detection is to detect the critical clouds which may cause escalation, if the leaks included in the HCR System would not present an explosion hazard, then the fixed gas detection system would not be expected to provide any actions. Even when we know the volume of gas which was released, this doesn't tell us whether the detection should detect the leak.

For example, if a major release occurred in an open process deck and there was an 80 mph wind blowing, the gas will be immediately dispersed and will not present the explosion risk that the fixed flammable gas detectors are expected to detect. Such information is not presented and therefore any assumptions or apportions of blame for poor detection rates on the placement of fixed flammable gas detectors from these leak detection statistics are immediately flawed.

Equally as important is the physical state of the detectors any at the day of a release. The assumption that the detection system is at 100% availability on the day of a release may also be fundamentally flawed.

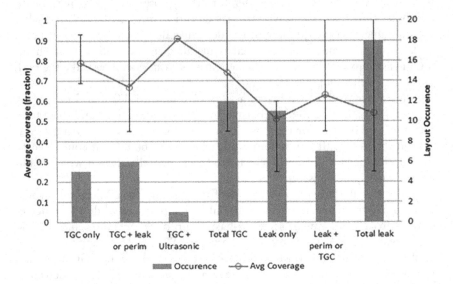

Figure 7.4 Breakdown of detection approach occurrence and detection fraction (7).

Figure 7.4 (8) shows the results of Hilditch et al.'s study on what methodologies have been applied in the UKCS and their corresponding coverage rates.

CFD Modelling in the Placement of Gas Detectors

Various CFD tools are available on the market today, and with the increased prevalence of gas detection mapping, the abilities of the modelling tools are being transposed across to modelling the effectiveness of gas detection systems. Within certain applications of design, these tools can be highly beneficial to the design engineer; however, care should always be taken to ensure that the limitations of the technique are fully understood, and that the specific type of CFD modelling is relevant (i.e. dispersion versus explosion analysis).

Typically, CFD tools are being used to model dispersion of flammable gas clouds in scenario-based designs, as previously discussed. Again, in certain circumstances, this can be of benefit, but of further benefit would be the application of CFD explosion analysis tools in the placement of gas detectors.

Bearing in mind the original design intent of flammable gas detectors is mitigating explosions, the dispersion of gas which does not pose an explosion risk is largely irrelevant, unless that location presents an ignition source and/ or an explosion risk (obviously, this is entirely dependent on the risk analysis and performance targets applied and is not applicable to all facilities).

The application of explosion analysis to review the explosion risk in an area will be of great benefit in setting the target gas cloud which can then be modelled in a gas mapping tool to verify detector performance against the said target gas cloud. This will allow true optimisation of the gas detection design without being dependent upon the environmental conditions which are known to be changeable to the point the results may not be reliable. Analysis of explosion potential in the area will be dependent upon the fixed structure and blockage which is less changeable on a day to day (or minute to minute) basis.

When considering the application of CFD modelling for the analysis of gas dispersion, caution must also be applied when considering the nature of the tool applied. Not all CFD packages are created equal and are not applicable for all applications. Equally there are dispersion simulation models which are technically not CFD tools, but rather based on simplified 2D Gaussian plumes, for example. Such tools are all useful if applied to the correct application with an understanding of the limitations.

For example, the 2D plume analysis may not account for near-field blockages and will provide a crude representation of gas dispersion. Such tools can be every bit as useful as a full CFD package, depending on what the designer is looking to determine.

The assumptions within any CFD modelling project must be considered, as with the limitations of the TGC method. While CFD tool providers will compete over who can run the greatest number of simulations in the shortest time, such arguments may not be as relevant as they appear. Such arguments may be useful sales tools, but for the engineer designing fixed flammable gas detection, other factors are more critical. Claims of running thousands of simulations can be impressive (and the technological advancement to do this *is* impressive), but are we even close to running enough scenarios to reflect the potential consequences of a release. While it is true that analysis of thousands of simulations eventually leads to a flattening of the effectiveness curve (as seen in Figure 7.5), this is functional of the limited simulation run. If the tool could analyse the trillions of scenarios which could occur, the effectiveness of the detectors being added would almost certainly change to the point that the design would change.

BS60080 presents an example figure which shows the number of recommended detectors levelling out after ~1,000 simulations. This could be used to justify a low number of simulations, but it is critical to note that the graph does not completely level off. The figure only goes to a maximum of 1,400 simulations. Consider that this figure keeps going to a realistic number of simulations—it is easy to consider that the number and location of detectors would be significantly different. The issue is one of scale.

While the convergence of a design will occur at a certain threshold of simulations, it is hard to argue that the design would not change when considering the true number of scenarios which are credible. This is a fundamental issue in the scenario-based methodology and shows that while the number of simulations is important, until we can run a truly reflective number of simulations, the design will continue to evolve. This also does not address whether the cloud migration presents an explosion risk in those areas, as previously discussed.

The argument therefore exists that if we cannot run a truly reflective number of simulations, why would a design which applies the TGC method, which is then validated by a small number of simulations to demonstrate credibility, not provide a more robust solution for most applications? While this would eliminate the back and forth arguments over who can run the greatest number of simulations, it may re-focus the efforts in gas detection back to what is fundamental in the application of fixed detection.

While the discussions and limitations around CFD dispersion analysis are relevant to external applications, when the designer is designing a system for internal applications, those limitations are greatly reduced.

In an internal and reasonably predictable environment, the application of CFD dispersion modelling is of great benefit. Such an analysis allows the modelling of HVAC-driven airflows, and while the release modelling still bears some uncertainty, the movement of species through the volume is more reliable. This provides detail on dead zones and likely locations of gas accumulation where gas detectors would be well placed. Such designs should, however, be suitable when the HVAC is both running and when it is switched off.

It is hard to move away from the argument that reducing detectors as a result of 'probability of detection' using either CFD or other scenario-based design could be dangerous from a systemic risk perspective. This could be argued to be a replication of the kind of design which led to Piper Alpha and many other disasters we have seen within the industry. The idea of design complexity simply means we do not understand something. This should be a warning when it is proposed that complex modelling be integrated in practice. Where this 'complex' (not well understood) approach is put into practice, we introduce the break in control and feedback between system controllers, such that failure becomes inevitable. The history of industrial accidents is a stark reminder of this, and as many have said before, those who ignore history are doomed to repeat it.

The uncertainty around both volumetric and scenario-based designs is significant. When considering, however, epistemic and aleatory uncertainty (11), this can help point in the direction of travel which would be best suited for the industry.

Considering epistemic uncertainty (uncertainty cause by a lack of data), this can be gathered by increasing data. With further testing of various blockage configurations and flammable gases, we could develop a database of target cloud sizes to be applied, thus improving the volumetric method. When considering aleatory uncertainty (intrinsic randomness of a phenomenon), this seems relevant to scenario-based design. With such seemingly random behaviour on a facility day to day in how a release will occur and behave, it becomes a significant task to reduce this uncertainty and provide a suitable number of simulations to account for this. This can therefore have a bearing in where precious resources should be placed in the development of methodologies to be applied.

The designer must be fully cognisant of these limitations when applying the tools, and the relevance of such modelling tools should always be considered. In the pursuit of improving methods of gas detection placement, the author believes CFD modelling for explosion analysis will bear the greatest and most relevant fruit for most applications with congested and confined areas. This application of CFD modelling bears the most resemblance to why we use fixed flammable gas detectors.

The Narrative of Gas Dispersion Modelling in Flammable Gas Detector Placement

Following from the discussion on CFD modelling in detector placement, the approach has been shown to be useful in some applications, but there exists a number of narratives associated with its application which require further research and peer review.

As previously mentioned, it is intuitive to believe that placement of gas detectors where you expect the gas to travel is a suitable approach. This can be true, for example, when designing a perimeter-based gas detection system. For general area hydrocarbon detection, however, it may not be the case.

Two prevalent narratives are discussed later with the need for further peer-reviewed research required.

Narrative 1: Using Gas Dispersion Modelling Will Produce a More Effective Gas Detection System: Is This True?

When the volumetric and scenario-based approaches are compared, the % coverage and % of scenarios detected are often directly compared. This often results in the inflated capability of the scenario-based method (12). These two metrics, however, should not be compared like for like. The reason for this is that they are targeting entirely different outcomes.

The volumetric method aims to detect clouds before they present an explosion hazard, therefore adding detectors through the volume will generally provide a reasonably uniform coverage contribution for each detector up to the point the system is 100% effective against a specific target cloud.

Scenario-based modelling, however, adds devices based on anticipated gas migration behaviour, often regardless of the explosion risk presented in that location. When monitoring device effectiveness as detectors is added, there will be a diminishing return based on the number of scenarios modelled. Picture, for example, an area where the prevailing wind direction is the influencing factor in cloud location. Once devices have been added to the location gas travels to, adding any more detectors will not return any added benefit as the scenarios show gas travelling to an area where detectors are already positioned. This therefore shows an effective system, requiring no further detection. This system has not, however, accounted for the areas where gas may migrate to in a smaller % of scenarios. Gas may still travel here in reality and may pose a significant explosion hazard. Arguably it could present an even greater explosion hazard than the area where the scenario-based approach has recommended detectors. The approach would therefore leave gaps in dangerous areas where the volumetric approach would place devices. It may also add detectors in an area of negligible risk, where a volumetric approach would not add any devices.

For this reason, any direct like for like comparison between effectiveness of scenario and volumetric approaches is misleading and philosophically flawed. This will be demonstrated in the discussion around narrative two.

Figure 7.5 (5) shows a comparison of how coverage is reviewed between volumetric and scenario-based detection methodologies. The primary difference between coverage is shown on the y-axis, with % coverage being the metric for volumetric coverage, and % of scenarios detected being the metric for scenario-based mapping.

In representing these graphs side by side, it can deceptively show that a comparable coverage is provided by fewer detectors when using the scenario-based approach. In particular, when validation studies test the effectiveness of both layouts using a scenario-based analysis, it creates a self-fulfilling prophecy. If detectors are placed on the basis of a theoretical simulation, then we use that theoretical simulation to test the effectiveness of the layout, it will present a highly effective system. It will also appear more effective than the volumetric approach. The problem is the validation of the scenario and volumetric-based approach validates the scenario-based design using scenario-based methods. If one were to validate a scenario-based design using a volumetric approach, the effectiveness of the scenario-based layout will be poor, as will be shown in narrative two, and is shown in the literature.

The difference is on the risk influencer (either frequency or consequence). The volumetric approach focuses on consequence and does not care whether the gas travels to a location in 99% of simulations, if the 1% presents a catastrophic hazard and the 99% presents no explosion risk. The focus is on the 1%. With scenario-based methods, the emphasis is on frequency, and on how many modelled scenarios will be detected. Acceptability criteria will set a minimum number of scenarios which have to be detected, with consequence bearing a lower importance.

Narrative 2: Scenario-Based Mapping Will Reduce the Number of Gas Detectors Required: Is This True?

The idea that scenario-based mapping will always reduce the detector count is an interesting narrative. It is interesting in light of the nature of the

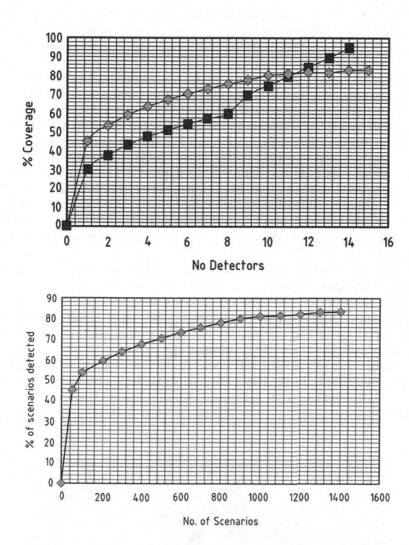

Figure 7.5 Geographic vs scenario gas detection effectiveness (11).

analysis—the detector count will likely be dependent upon the number of scenarios run and the acceptance criteria.

If we take an extreme where a single scenario is run, and we place a single detector where the gas goes, this will provide a 100% effective system with a single device.

We can also take a second extreme where we run trillions of scenarios and find that the gas can go anywhere in the area. This would essentially result in detectors being placed throughout the volume, increasing the number of detectors.

These extremes are unrealistic in practice; therefore, a comparison must be made with published literature showing a scenario-based layout (13) to gain a more accurate understanding of the level of optimisation capable with the approach.

The initial design presented in the literature is a representation of 5 m spacing using only point gas detectors. This therefore informs us that the cloud which will present a damaging cloud is 6 m in diameter (if the approach has been implemented in accordance with the underlying philosophy).

The initial consideration in reviewing a 5 m spacing approach using only point gas detectors throughout the entire volume is somewhat misrepresentative of a realistic application of the approach. A performance-based approach to the method allows relaxation of the target cloud in areas which are more open, and also does not strictly require 100% 5 m spacing through the area. The approach also allows the user to apply modern gas detection technology in detecting the clouds which does not appear present in the example.

Figures 7.6[1] and 7.7 show an example misrepresentation of an adequate real-world volumetric design.

If we are to run a comparison between design strategies, we consider the aforementioned study which presents a reduction of 22 detectors to 11 detectors by modelling over 11,000 dispersion plumes, while claiming that coverage improves to over 80% detection (13).

Another way of phrasing this coverage improvement, however, is that only 80% of the leaks simulated can reliably be detected. This is a change from a design which, while being over engineered, would detect all clouds which present an explosion risk. What is not clear from the study is if these

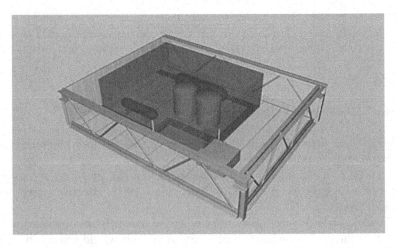

Figure 7.6 Volume where an accumulation of gas would present an explosion overpressure of concern if ignited.

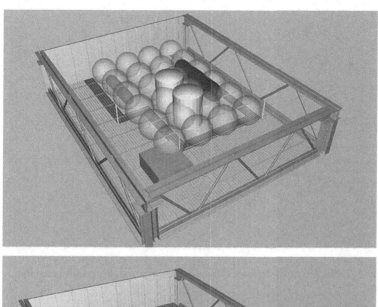

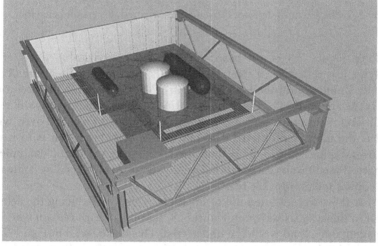

Figure 7.7 Misrepresentation of an adequate volumetric approach.

80% of detectable leaks present an explosion hazard, or if the 20% of undetected leaks will present a risk of explosion.

To compare the design strategies, we consider again that with the 5 m spacing applied, we can assume that a 6 m cloud in the volume will present an explosion hazard. We therefore run an analysis of what volumetric coverage is achieved from a layout which reduces 22 detectors to 11.

Figure 7.8 shows a similar reduction in detection from the layout presented in Figure 7.7 which could be a result of a scenario-based dispersion analysis.

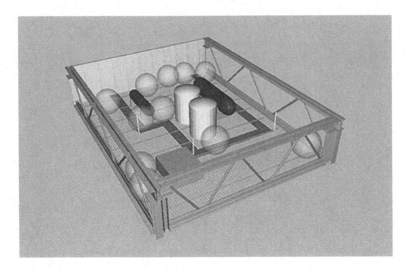

Figure 7.8 Placement of detectors based on anticipated dispersion behaviour.

While Figure 7.8 is not based on a specific scenario-based dispersion analysis, it is reflective of the type of results found in the literature (13). A specific simulation of the layout in Figure 7.8 has intentionally been omitted as the application of different scenario-based dispersion tools will result in different detection layouts, potentially leading to debate over which tool is the most appropriate. The purpose of this discussion is to review the overarching philosophy behind the methodology. Specific examples of scenario-based results can be found in the literature produced by companies developing such tools (12, 13).

What these studies often fail to highlight is that by reducing the detector count in this way, as shown in Figure 7.8, significant gaps remain where the damaging 6 m cloud would be undetected. This can be seen in Figure 7.9.

Figure 7.9 demonstrates the vast reduction in coverage of potentially dangerous cloud accumulation. By inference, this means that a gas cloud which would result in an explosion could occur undetected in the vast majority of the volume when using the scenario-based approach.

A designer and facility owner must therefore decide if they will select a design methodology which will leave such a significant undetected volume based on a finite run of leak simulations. While a credible worst-case scenario can be determined in certain applications, such a Fire Safety Engineering in building design, the application of such tools in external, high-hazard industries, is another matter entirely. Careful selection of the approach is therefore critical in the process of setting performance targets.

Additionally, the literature often omits the application of modern technology in the designs of volumetric detection. Implementing OPGDs instead of

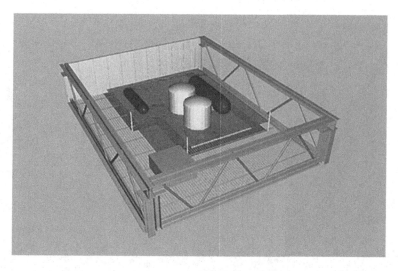

Figure 7.9 Coverage gaps to the dangerous gas cloud using the scenario-based approach.

point type devices can drastically reduce the device count, and increase the effectiveness of a volumetric design.

Applying the volumetric grades from Figure 7.6, along with modern technology presents arguably the most optimised and safe design, with almost 100% coverage achieved using only four devices. This is a significant improvement for both optimisation and safety over the previously shown 11 point detectors providing only 80% coverage from the scenario-based example.

Figure 7.10 shows an example solution using modern technology. Note that this solution may not be credible in all circumstances, and based on the nature of the gas hazard present, a combination of OPGDs and IRPGDs may be required.

The demonstration given earlier shows that a volumetric approach using modern technology provides potential for the most optimised and safe solution *in this specific application*. This may not always be the case and is critical when deciding what performance targets to set. In this case, it can be reasonably well assumed that if the 6 m cloud which presents an explosion risk is present in the volume, it will be detected.

As a final note, the literature demonstrates that in the event of an accident which is then demonstrated to have been foreseeable, where steps were not taken to mitigate, this leaves companies in a dangerous predicament in the eyes of investigators, and in some cases, the law (15). Can designers rest easy while removing detectors based on such limited samples with such significant potential consequences?

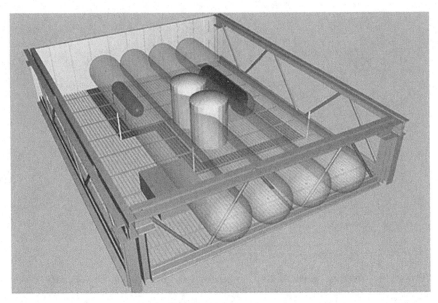

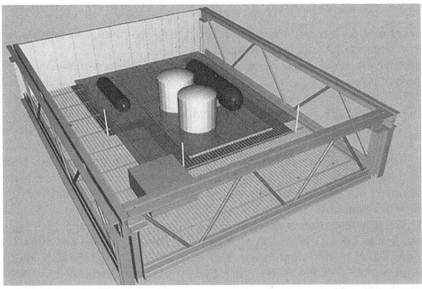

Figure 7.10 Optimised gas detection layout applying the volumetric approach.

Note

1. Figures 7.6–7.10 are produced using HazMap3D [14. Milne D, McNay J. HazMap3D, 3D F&G Mapping Software. Micropack (Engineering) Ltd.; 2016.] with permission from Micropack (Engineering) Ltd.

References

1. Hitchens C, Dawkins R, Harris S, Dennett D. The Four Horsement: The Conversation That Sparked an Atheist Revolution. Random House; 2007.
2. HSE. Offshore Technology Report OTO 93 002: Offshore Gas Detector Siting Criterion Investigation of Detector Spacing. Health and Safety Executive; 1993.
3. Kelsey A, Ivings MJ, Hemingway MA, Walsh PT, Connolly S. Sensitivity studies of offshore gas detector networks based on experimental simulations of high pressure gas releases. Process Safety and Environmental Protection. 2005; 83(3):262–9.
4. Pittman W, McNay J. The Case for the Target Gas Cloud Approach in Gas Detector Placement. Available from: https://f.hubspotusercontent20.net/hubfs/9039428/Micropack_Fire_And_Gas_January_2021/Pdf/TN-The_Case_for_the_Target_Gas_Cloud_Approach_in_Gas_Detector_Placement_Rev_0.1.pdf.
5. BSI. BS60080 Explosive and Toxic Atmospheres: Hazard Detection Mapping—Guidance on the Placement of Permanently Installed Flame and Gas Detection Devices Using Software Tools and Other Techniques. BSI Standards Limited; 2020.
6. Milne D, Davidson I. Flame Detection Assessment Gas Detection Assessment (FDAGDA). Micropack.
7. Solutions SG, Technology B. Gas Buildup from High Pressure Natural Gas Releases in Naturally Ventilated Offshore Modules. Joint Industry Project; 2000.
8. Hilditch R, McNay J. Addressing the Problem of Poor Gas Leak Detection Rates on UK Offshore Platforms. In: Proceedings of the Ninth International Seminar on Fire and Explosion Hazards. Vol. 2: 21–26 April 2019. Saint Petersburg, Russia; 2019.
9. HSE. Offshore Hydrocarbon Releases 1992–2016. Health and Safety Executive; 2016.
10. McGillivray A, Hare J. Research Report 672 — Offshore Hydrocarbon Releases 2001–2008. Prepared by the Health and Safety Laboratory for the Health and Safety Executive; 2008.
11. BSI. PD7974–7: 2019 Application of Fire Safety Engineering Principles to the Design of Buildings, Part 7: Probabilistic Risk Assessment. BSI Standards Limited; 2019.
12. DNVGL. Scenario-based Fire & Gas Mapping as a Way to Optimise Detection Layouts. In: Presented at FABIG TM942018. Available from: https://www.fabig.com/publications-and-videos/technical-presentations-videos/technical-meeting-094/.
13. Marszal E. The Case for Scenario Coverage for Gas Detector Placement. Kenexis; 2019.
14. Milne D, McNay J. HazMap3D, 3D F&G Mapping Software. Micropack (Engineering) Ltd.; 2016.
15. Andrews G. Was it foreseeable? Lecture to UKELG University of Leeds; 2016.

8 Notes on Specialised F&G Hazards

The provision of F&G detectors often consists of more than simply process area flame and flammable gas detectors. Additional considerations for the F&G system include HVAC systems, ultrasonic/acoustic detection design, and toxic gas detection design. Traditionally, these ancillary design considerations/technologies have even less guidance and standardisation around their application.

Picture a study in which an offshore facility which presents a toxic gas hazard is undergoing review to ensure that a sufficient number of toxic gas detectors are placed. The hazard exists at locations where personnel would likely be present throughout the area (more on specific toxic design strategy later). During a design meeting, the proposed number of detectors to meet the agreed performance targets is deemed excessive by an onshore-based representative of the operator. Engineers in the room query that if the proposed design is excessive, how many would be acceptable and who would sign off on such a reduction in detectors. With limited guidance on what an acceptable number of devices would look like, an operator's response of 'just buy 10', while cavalier and likely unsafe and unjustifiable, may not be unusual.

In any particular application, ten detectors may well be suitable, but where is the engineering justification? Unfortunately, with traditionally such limited guidance and standardisation on the subject there may be no way to disprove of the suitability of 'just getting 10'. I would wonder in such cases like this, if onshore representatives had to work and live on the facility in question, would the approach be quite as haughty? In fact, with a subject like F&G design where there is rarely a right or wrong answer on coverage adequacy I often consider 'if I had to stay on this facility would I be comfortable with the design?' A subjective criterion, yes, but in the absence of a prescriptive design standard, such performance-based decisions, with the all-important engineering justification of course, are critical. Fortunately, this issue should be less prevalent today with the introduction of BS60080 (1).

DOI: 10.1201/9781003246725-8

HVAC/Air Inlet Detection

The philosophy behind HVAC, or air intake, gas detection is to prevent flammable gas from a hazardous area being ingested into an area classified as non-hazardous and reaching a potential ignition source. There are consequently multiple components which are important in the design of a gas detection system to prevent against such an outcome.

The gas must be detected, and an action must be initiated which will stop gas from flowing through the duct to reach an ignition source. This should all take place before gas has the opportunity to enter the first non-hazardous space. An effective detection positioning can be seen in Figure 8.1. An example of an ineffective detector positioning which could not ensure detection before gas reaching an ignition source is shown in Figure 8.2.

In order to make sure that gas is detected and prevented from further ingestion before an ignition can occur, we should know, or calculate, the following:

- What is transit time from Point A to Point B as per Figures 8.1 and 8.2?
- What is response time of the detection system including the:

  - Gas detector
  - Control equipment
  - Fire damper/intake closure

The response time of the system should be less than the transit time to create a safe system.

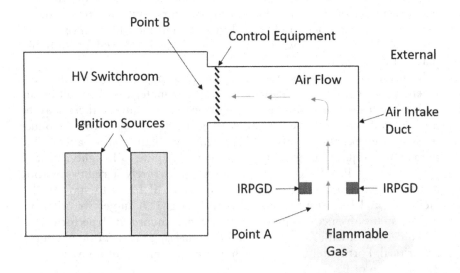

Figure 8.1 Effective HVAC gas detection placement.

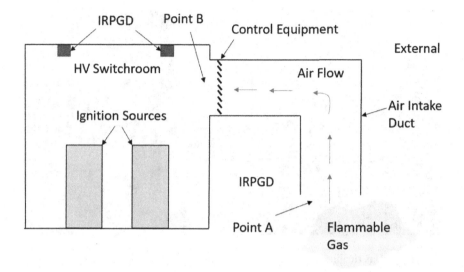

Figure 8.2 Ineffective HVAC gas detection placement.

Air flow systems which will require gas detection are numerous and not simply limited to non-hazardous space HVAC systems. Examples can also include intakes to machinery spaces, the machinery itself (for example, a combustion engine air intake), and turbines. Figure 8.3 (2) shows a gas turbine with the four 'ducts' indicated.

In this type of application, gas detection would be required on the following:

- the combustion air intake to prevent ingress of gas to the combustion process potentially causing turbine overspeed and a spark;
- the ventilation intake to prevent ingestion of gas which may contact the internal hot surfaces of the turbine;
- the ventilation exhaust to detect any potential gas release from within the turbine.

The control actions for each of these ducts can differ. For example, detecting gas at the combustion intake would typically require shutdown of the intake and also the turbine, whereas with detection at the ventilation exhaust, often the turbine process is shutdown to reduce ignition potential and release of further gas (if the leak has originated within the enclosure), but the ventilation remains running to flush the gas from the turbine's internal congested environment. For obvious reasons, the combustion exhaust does not require gas detection.

With respect to how and where we should place the detectors in the ducts, the UK HSE provides useful guidance on the topic (3). The philosophy behind gas detection in air intakes requires:

Figure 8.3 Gas turbine duct representation (2).

- detectors be placed as close to the face of the duct as possible (the side of the duct closest to the hazardous area to maximise response time of the system);
- detectors should be placed to maximise the likelihood of detection of non-uniform distributions of gas (i.e. all detectors shouldn't be placed on one side of the duct if the duct is suitably wide and gas could by-pass the devices);
- the detectors should achieve a fast response to the presence of gas (often a T90 time of ~2 seconds).

While not specified in the HSE guidance, a useful rule of thumb to achieve detection of non-uniform distributions of gas relates to the technology applied. The following technology can be used on the basis of the width of the duct in question (bearing in mind that the detector manufacturer should always be consulted to ensure that the detector is suitable to the application and target gas):

- <650 mm width—fast response IRPGD
- 650–1300 mm width—extended path point IR
- >1300 mm—duct mounted OPGD

It is important to note that each duct is unique and will require specific consideration. Generally, devices should be mounted on different sides of the duct as far as possible, and a sufficient number of devices should be installed to ensure suitable coverage in the duct. This should also ensure control actions can be successfully implemented as gas travels across the duct.

Ultrasonic/Acoustic Design

In the event of a pressurised gas leak, the leak emits ultrasonic energy. The application of acoustic/ultrasonic gas detectors detects this ultrasonic energy from the release. These devices do not require a direct line of sight and will operate even when the gas is dispersed meaning that detection will be successful even if the environmental conditions and leak behaviour do not allow a dangerous gas accumulation to form. Such device effectiveness depends, however, on gas flow rates, hole size, and pressure of the stream. The effectiveness is also heavily influenced by the background ultrasonic noise level. As such, a background mapping study is generally required (where possible) to verify the background noise level and determine the effective detection envelope of the devices. This is critical in the mapping and placement of such devices.

The travelling ultrasonic energy can be impacted by blockages and confinement in a similar way in which a flame detector cannot repeatedly and reliably 'see' around corners. It is not impacted to the same extent, however. The comparison is a simple one. You cannot see around corners. While you can still hear around a corner, the noise can be muffled by the obstruction. Such factors should be considered in a suitable ultrasonic/acoustic gas detection design.

One of the primary benefits gained from the application of ultrasonic gas detection is the speed of response from the moment of release. If the ultrasonic energy generated from the release is sufficient, detection can be near instantaneous (dependent upon any time delays applied in the detection to prevent false alarm). The technology provides the benefit of not having to wait until the gas has accumulated before detecting the release.

With a volumetric gas detection design, for example, the gas cloud has to accumulate and approach the state where it begins to present an explosion hazard before detection occurs.

Similarly, with a scenario-based flammable gas detection layout, to achieve sufficient coverage through the area to detect leaks as quickly as an ultrasonic detector, the designer would likely need to install an unacceptable number of detectors. The ultrasonic detection, however, provides a reasonably large detection footprint which would mean that this over-engineered scenario-based solution is not required.

With reference to this large detection footprint, some modern devices are capable of a detection radius up to 20 m (4). This allows the designer, dependent on obstructions, background noise level, and release conditions, to potentially design a sufficient system with the strategic placement of a few devices which will cover a significantly large area.

With the strategic placement of the devices, ultrasonic gas detection does not suffer, to the same extent, from the issues discussed in Chapter 2, Mitigation Actions. This is where gas may be detected in one area, despite having emerged from another. As previously discussed, this can result in ineffective mitigation actions in the wrong area. Ultrasonic gas detectors can be more easily designed to detect releases specifically within their area of operation.

As with most technology, there are limitations which must be addressed during design. One of the primary limitations is the inability of the technology to distinguish between pressurised leaks of different aerosols. For example, the high-pressure release of instrument air cannot be distinguished over a pressurised release of Methane. This presents the possibility of alarm and subsequent mitigation action to the release of a non-hazardous material. This limitation can be mitigated by introducing a time delay on the device such that normal operational 'hisses' of non-hazardous gas can be discounted.

Another limitation is the variance in detection capability based on the background noise level. This presents a possible change management issue that the effectiveness of the detection will vary through the facility lifecycle. This can, however, also be designed into the system. This can be done in two ways.

The first is to ensure that the application of an appropriate change management system to verify and revalidate detection capability as the operational condition in the area of concern changes. This can be coupled with regular audit of the background noise to ensure the system remains adequate.

The second process could be to assume a worst-case scenario background noise level from the start of design. This can, however, result in unnecessary over engineering of the detection layout.

Another potential limitation of the technology is the relatively unknown performance characteristics to varying stream compositions. The devices respond well to gas streams of 100% gas composition; however, as moisture/liquid is introduced to the stream, the sound pressure level (SPL) rapidly reduces upon release, which reduces the effectiveness of ultrasonic detection.

While this limitation can be designed out by solely applying the devices in gas processing areas, the likelihood that moisture will be present somewhere in the stream within the area is high, and as the designer does not know the point at which the moisture will begin to drastically reduce detection capability (i.e. the percentage of the stream of liquid at which point the ultrasonic detection becomes ineffective), this can be challenging to fully account for.

Given these strengths and limitations, there is value to be gained in certain facilities and F&G designs from the use of ultrasonic detection. The application of performance-based principles is critical in their application. For example, high-risk areas processing high-pressure flammable gas with limited moisture in the stream will benefit from the fast response ultrasonic detection. Equally, when carrying out the performance-based gas detection design, if the environment would suggest that flammable gas detection is unlikely to be effective as the gas is unlikely to accumulate local to the release (e.g. a vastly open onshore pumping station exposed to severe weather), the local detection achieved by ultrasonic detection can provide benefit.

The devices can also be used to supplement a traditional flammable gas detection system. For example, where a volumetric IR-based gas detection system is installed, ultrasonic detectors can supplement the detection by being positioned strategically in high-risk areas. This will therefore provide a suitable detection system in detection of dangerous clouds and providing executive actions (from the volumetric IR system), and a fast response alarm system which can alert operators to a potentially developing hazardous condition (from the Ultrasonic detectors). The designer may also decide to cross vote the IR devices with the ultrasonic devices to enhance executive action reliability. This can also be assumed to be robust as it is unlikely a single false alarm source will cause both diverse detection technologies to alarm at the same time.

Crucially, as with any F&G technology being designed into the layout, the designer must consider what the device performance will provide, specifically in the environment it is being applied to. This is likely to require liaising with the device manufacturer to fully understand the detector capability and the specific facility and stream composition in hand. The designer must also consider, along with the facility owner, the likelihood of the stream pressure dropping over the lifecycle of the facility to the point where the ultrasonic detectors become ineffective. This is a likelihood when wells approach the end of their life, where the pressures have reduced well below the original design criteria.

An area of design which requires further consideration by the industry is the relevance of the design strategy of simply engulfing the area in the anticipated detector footprint of coverage. At the time of writing, further research would be well served to investigate the suitability of placing the cone of coverage in the area with limited anticipation of leak pressure, orifice size, and stream composition. Even more critical is the full impact the environment's obstructions make on the detection efficiency of the detectors.

One final point which can present an issue in design is where a background noise level survey cannot be completed. This survey is crucial in applying an accurate representation of the detector's footprint.

Consider human hearing, for example. You may not be able to hear the person sitting next to you because you are in a noisy room. In a quiet room, however, you can hear someone clearly at a greater distance from you. High background ultrasonic noise levels will reduce the detection capability of ultrasonic detectors in the same way.

Where a facility is yet to be constructed, this survey cannot be carried out, and assumptions therefore need to be made. The designer must make an informed decision as to how conservative to be in such cases, and these assumptions should always be validated on site when the facility becomes operational. While the designer will not want to be overly conservative during design, it is pertinent to be cognisant of the significant expense of adding detectors when a facility has become operational. It may be safer to err on the side of caution during design to make sure changes are not required after commissioning of the site.

Toxic Gas Detection

Unlike the design of a flammable gas detection system, where the primary detection target is typically not to protect individuals from small releases of flammable gas, a toxic gas detection system is specifically intended to detect small releases which may have a high toxicity and present an imminent danger to anyone within the volume.

Toxic gases such as hydrogen sulphide (H_2S) are present on many oil and gas facilities and can range from relatively low levels (up to 500 ppm) to facilities with an extreme H_2S risk (~30,000 ppm up to where it begins to register as a % of the stream). Such toxic gases become hazardous to health at very low concentrations; for example, H_2S can cause damage to the lungs at approximately 50–100 ppm and can cause rapid unconsciousness at approximately 700–1,000 ppm (5). In order to determine the risk posed by various toxic gases, workplace exposure limits are specified, for example, in Europe in EH40 (6). These, and other guideline limits, set values of exposure across various timelines in which humans can be safely exposed. For example, time weighted average (TWA) and short-term exposure limit (STEL) refer to the limit of safe exposure after 8 hours, and 15 minutes of exposure, respectively. Other values applied globally include Emergency Response Planning Guidelines (ERPG) and Acute Exposure Level Guidelines (AEGL) which apply similar logic in their application.

Unless the anticipated gas behaviour and application suggest that the gas will settle at low levels or high levels, fixed toxic gas detection, as with portable gas detection, is generally placed within the breathing zone of people. The devices are also generally placed along walkways, escape routes, entrances, and exits. This is intended to detect toxic gases in the locations where personnel may be present who are operating outside of any permit to work system (those who are following a permit to work type scheme will presumably be following additional safety protocols when working for an extended period in an area where toxic gas release is credible).

Where toxic gas is a potential risk, this does not always necessitate the application of a toxic specific detection system. When dealing with the potential presence of a toxic gas, depending on the application, the concentration of the stream, and the set points of the detectors, some operators apply the approach of inferring a toxic gas release through the flammable gas detection system in the area.

For example, where fast response IR gas detection is installed in an area (at relevant locations), and the set point for alarm is suitably low, these detectors may alarm to the presence of flammable gas in the stream, before a toxic gas detector would alarm to the low toxic concentration in the volume. Such a delayed response may be a result of the limitations in response time of the traditional detection technologies applied in the detection of toxic gases (i.e. electrochemical and semiconductor-based technologies discussed in Chapter 6).

A limit often applied for H_2S, for example, is where the stream contains at least 2,000 times the flammable gas concentration to the toxic concentration. Continuing with H_2S as an example, this would mean that, for every one part of H_2S, there are 2,000 parts flammable gas. This equates to 500 ppm of the stream being H_2S (note this example is only applicable to H_2S).

Where the IR flammable gas detectors are set to alarm at 20% LEL, when we take 100% LEL of Methane to be ~5% flammable gas in the total volume (this is often also cited as 4.4% Methane in air dependant on the resource), we can assume that an alarm is achieved at 1% Methane in air (20% of the LEL value = 1% of the total volume). This means that the operator infers detection of 1% of the volume in air, and where the concentration in a stream is 500 ppm, this infers detection of 5 ppm H_2S in air. Five ppm of H_2S is often cited as the limit of acceptable exposure (i.e. the TWA), which informs the decision to infer the presence of H_2S in an area presenting a risk up to 500 ppm H_2S in the stream.

Again, lessons can be learned in this area of F&G design from Fire Safety Engineering. PD7974–6 (7) discusses the importance of monitoring effluents of fire in the breathing zone, and any zone where occupants would move through. The approach discussed in PD7974–6 emphasises the importance of available safe escape time (ASET) and required safe escape time (RSET) calculations. The ASET is the time occupants have to reach a place of ultimate safety before the location becomes unsafe, with RSET representing the calculated time it will take for occupants to escape. Fundamentally, a design is said to be suitable where ASET > RSET.

This approach is the basis of the toxic gas detection design philosophy in BS60080.

In practice, this means that designers are well served to focus on two factors in fixed toxic gas detection design. The first is the anticipated location of personnel, and the second is the ease of evacuation from the area.

When considering where personnel are expected to be present, the designer should account for personnel carrying out work in the area, and also any personnel who may simply be passing through the area. The potential pre-alarm and pre-escape behaviours will vary between these occupant characteristics.

Personnel who are working in an area with a sufficient risk of toxic exposure should fall under a permit to work system where they are familiar with the toxic risk in the area and the safety controls in place (potable toxic gas detectors, gas sampling in the area, awareness of prevailing wind direction, etc.). Occupants passing through the area may not have this level of detail on the toxic risk but must be protected all the same.

Essentially, this results in fixed toxic gas detection designs which place detectors on escape routes, regularly used walkways, entrances and exits, and usually located within the occupant breathing zone. These devices are often positioned at a nominal spacing agreed with the operators, which can be dictated by the ease with which occupants can escape (is it a simple progressive horizontal evacuation or are stairs/ladders involved? etc.).

This design philosophy closely resembles the performance-based volumetric approach; however, scenario-based methods can also be used here in understanding possible cloud dispersion behaviours. Just as important, however, would be evacuation modelling in understanding the RSET. Within toxic gas detection design, this form of analysis is, however, rarely, if ever, carried out. It is the belief of the author this would present a significant improvement in the holistic safety from toxic gas release.

While using probabilistic analysis to model where the toxic gas may spread and accumulate, the designer must still be aware of where the occupants will move through the area, and make sure that they are not moving into a hazard at breathing height. Simulations can provide a view of how a toxic gas release can behave, and allow the designer to position detectors according to this anticipated cloud movement.

It must be noted that carrying out such an analysis in an external area presents the same limitations and issues as scenario-based flammable gas detection design in the same type of application. Care should therefore be taken if applying such a strategy. In practice, designers may feel that with such an immediate risk to personnel in the event of a toxic gas release, positioning of detectors based on anticipated cloud movement may not provide the same level of safety as assuming the cloud could affect any personnel in the area of release, regardless of environmental factors at the specific moment of release. That said, scenario-based approaches can be useful in determining the 'worst-case' extents of a toxic release, e.g. in large facilities, we can model how far a toxic gas cloud could be likely to travel.

Ultimately designers must review the specific risk and specific application and take a view on what the most appropriate approach is for that application.

It is also crucial to note that fixed toxic gas detection should never solely be relied on to protect personnel against a toxic gas release. Safe operating practices are always critical. Such factors can include portable gas detection, breathing apparatus, and safe operating practices (i.e. permit to work, restricted access).

References

1. BSI. BS60080 Explosive and Toxic Atmospheres: Hazard Detection Mapping—Guidance on the Placement of Permanently Installed Flame and Gas Detection Devices Using Software Tools and Other Techniques. BSI Standards Limited; 2020.
2. McNay J, Duncan G, Davidson I, Knox M, Dysart C. Micropack Fire and Gas Detection Design and Technology Course. Micropack (Engineering) Ltd.; 2020.

3. HSE. HSE Information Sheet, Advice on Gas Detection Strategies for HVAC Duct Inlets, Offshore Information Sheet No. 5. 2008.
4. Emerson. Incus Ultrasonic Gas Leak Detector Product Data Sheet. Emerson Automation Solutions; 2017.
5. OSHA. Hydrogen Sulphide Health Hazards. Available from: https://www.osha.gov/hydrogen-sulfide/hazards.
6. HSE. EH40/2005 Workplace Exposure Limits. TSO (The Stationery Office); 2020.
7. BSI. PD7974–6:2019 Application of fire Safety Engineering Principles to the Design of Buildings Part 6: Human Factors: Life Safety Strategies—Occupant Evacuation, Behaviour and Condition (Sub-system 6). BSI Standards Limited; 2019.

9 Competence

In some ways, the anecdotal story to kick off this chapter is the easiest of all, and is very similar in some respects to the hypothetical in Chapter 5.

It pertains to the promotion of services or products relating to the subject of F&G mapping, and the potential that exists to sell based on the limited understanding and research into the subject matter. Clearly, a possible barrier to selling software for F&G mapping, for example, is that those you are selling it to may feel that they do not maintain the correct level of ability to use the tool effectively, and may prefer to outsource this to a competent third party. Rather than moving to provide education through training and support of the subject matter, thereby improving the overall quality of knowledge on the subject matter, it could be far quicker and easier for the provider to simply state that one does not need to understand how a fire or gas detector works in order to do F&G mapping, you simply need the software tool to do the design for you. I would hope that the forerunning chapters in this book would exemplify why this is not only untrue, but remarkably dangerous.

This brings home the importance of not only competence and experience, but *relevant* competence and experience.

The setting of performance targets and acceptance criteria, and the analysis of what 'adequate' looks like have been a grey area for decades. Practitioners have spent their entire career in pursuit of improving the methods and tools which allow designers to further justify a performance-based layout.

The nature of a numerical acceptability threshold is a contentious area. When we consider the uncertainty around mapping processes and technology in the field, the idea of a nominal percentage at which point a design becomes adequate is clearly not an optimal solution, but it may be a necessary one. Some in the industry perhaps reasonably have fought against the idea of including target percentages of coverage in design philosophies; however, for this to be useful in practice, competence in analysing coverage maps is even more critical.

The effort towards achieving consistency in detection layouts may be a driver behind incorporation of target coverage factors, and while this is an understandable pursuit, the effort may be misplaced when considering the

DOI: 10.1201/9781003246725-9

performance-based nature of F&G design. While having target percentages is widespread, some believe that it may promote behaviours of simply aiming to achieve that percentage, rather than designing to optimise and ensure safety (1). The argument exists that the target percentage is there to protect the designer rather than the facility.

An emerging danger is that some believe software alone to be the solution to the problem—simply import a model and let the tool place the detectors. This puts an exorbitant amount of trust in the developers of such tools. It was said best by T. Cuyler Young in 'Man in Nature' that

> we pretend that technology, our technology, is something of a life force, a will, and a thrust of its own, on which we can blame all, with which we can explain all, and in the end by means of which we can excuse ourselves.

The debate is unlikely to be settled any time soon regarding inclusion of target percentages; however, the debate cannot lose sight of the philosophy behind detection design. Discussions today focus primarily around how to generate the target percentage rather than whether it is an acceptable metric to use. Considering the traditional methods of design, those involved from that early stage had experience of field-based performance, having been experts in the development of technologies, the installation, commissioning, and maintenance of the devices (2).

There was also a great deal of investment, both time and financial/resource based, in testing of technologies and explosion/fire/gas cloud behaviour. These efforts and experiences led to a pragmatic focus when it came to detector positioning, only with the absence of the tools available at our disposal today. This is not to suggest that the problem was solved in those early days. Quite the contrary, there is much work to be done, but there is wisdom in those early efforts, and they are the shoulder from which we should all perch.

When the term F&G mapping is mentioned now, thoughts immediately go to a software-based analysis (3), with some, as previously discussed, believing that a detection layout can adequately be designed solely by the application of a software tool. This reliance on black box software analysis rather than engineering first principles can heavily rely on assumed environmental and release/fire behaviours, with theoretical detection capabilities unlikely to be experienced in the field (4).

This brings us to the area of relevant competence, an issue which is not unique to F&G design. The National Research Council's NAE report on Macondo (5) states 'There are few industry standards for the level of education and training required for a particular job in drilling'. For those in the safety industry, it is arguable that this is not solely an issue in Drilling. Furthermore, the US Chemical Safety and Hazard Investigation Board (6) states 'the Macondo incident almost automatically raises questions about competency of the personnel involved'. Such statements which dovetail so gracefully with

F&G mapping show a systemic issue in O&G safety, particularly around the niche areas of active protection. A holistic shift from 'how many years have you worked in safety?' towards 'what experience do you have specifically in the design and development of F&G Systems' may assist in improving detection designs across the board. In the same way, one would not employ an electrician to do your plumbing, an expert in quantitative risk analysis (QRA) does not necessarily have expertise in F&G detection system design.

Safety First

Within many industries, financial pressures are a primary driver behind behaviour. In times of plenty, budgets can become bloated which leads to genuine practice of safety first. The potential for behavioural change when markets crash and financial pressures emerge is all too well documented in most major accident investigations.

The drive towards optimisation to allow new projects to be sanctioned in tough market conditions is tempting in the board room. This creates a useful gap in the market for software tools which can 'do everything with minimal engineering input'.

While on the face of it, this is good news for industry, and this becomes problematic considering the lack of any peer-reviewed scientific evidence, at the time of writing, to suggest that it is a suitable method of design, or even that it results in reducing detector counts as is so often claimed. In fact, practice suggests the contrary.

Consider the discussion in Chapter 2 relating to target fire size. It can be tempting for operators to increase the target fire size to the anticipated fire in the area, but this misses the philosophical requirement of flame detectors. As previously discussed, the designer runs the risk of losing the mitigative function by waiting until the fire is too large to be controlled, and secondly, the design will rely on devices detecting fires they are not certified to detect.

Verweijen et al. (7) state, 'For oil companies, maintaining the flexibility to adapt practices to organizational needs is a deeply embedded institutional norm. Attempts for standardization of practices is therefore generally resisted'. It is credible therefore that lessons learned in a particular company can be discarded as a result of this required flexibility and as the market conditions dictate. When the fixes to those lessons learned are discarded, we open up the possibility (or certainty) for the same mistakes to happen again.

Verweijen continues:

the pervasive variability in training also is driven by the 'boom-and bust' cycle in the oil industry. Periods of high investments followed by periods of underinvestment have created chronic discontinuities in experience and competence across the pool of industry workers.

Figure 9.1 (2) shows an example lifecycle of safety consideration.

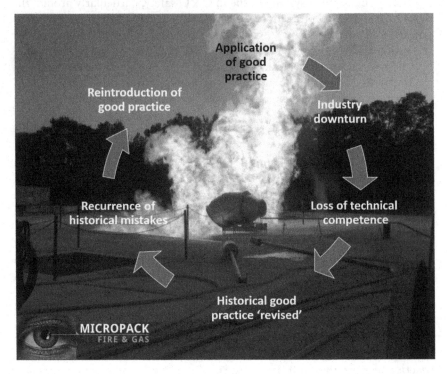

Figure 9.1 Lifecycle of technical practices (8).

In summary, the lessons learned must live longer in the memory, and any developments and improvements in the field of F&G mapping and safety must start from the point of targeting what 'safety' looks like for the specific application. The development of facilities which appear to be the same as those which have come before should be scrutinised in light of any apparently subtle differences, which could have a significant safety impact. Those facilities which are novel with new processes or equipment should be analysed using performance-based principles to determine if the preventive and mitigative functions for the system are fit for purpose.

Not by automatically running simulations and throwing probability calculations at every site, but only from taking this step back to ensure that we are applying appropriate risk models and technology, can we move towards truly performance-based and ultimately reliable F&G system.

References

1. McNay J. Desensitisation of optical based flame detection in harsh offshore environments. International Fire Professional. 2014;(9).

2. McNay J. Competency in F&G mapping. International Fire Protection Magazine. 2019.
3. Milne D, McNay J. HazMap3D, 3D F&G Mapping Software. Micropack (Engineering) Ltd.; 2016.
4. McNay J. The Role of Engineering Judgement in Fire and Gas (F&G) Mapping. International Society of Automation; 2017. Available from: www.isa.org/Safety-and-Security-Division/FG-June_2017/.
5. Council NR. Macondo Well Deepwater Horizon Blowout: Lessons Learned for Improving Offshore Drilling Safety. Washington, DC: The National Academies Press; 2012.
6. CSB. Investigation Report Drilling Rig Explosion and Fire at the Macondo Well. US Chemical Safety and Hazard Investigation Board; 2016.
7. Verweijen B, Lauche K. How many blowouts does it take to learn the lessons? An institutional perspective on disaster development. Safety Science. 2019;111:111–18.

Index

Printed in the United States
by Baker & Taylor Publisher Services